W9-BJM-589

Humans in Space

21st Century Frontiers

Humans in Space

21st Century Frontiers

Harry L. Shipman

Plenum Press • New York and London

Library of Congress Cataloging in Publication Data

Shipman, Harry L.
Humans in space: 21st century frontiers / Harry L. Shipman.
p. cm.
Bibliography: p.
Includes index.
ISBN 0-306-43171-8
1. Astronautics. 2. Outer space—Exploration. I Title.
TL790.S488 1989 88-33640
333.9′4—dc19 CIP

Plenum Press is a Division of Plenum Publishing Corporation
233 Spring Street, New York, N.Y. 10013

Printed in the United States of America

Preface

We all think of the space program as one of the most forward-looking of human activities. Meeting technological challenges, exploring the unknown, and the lure of the next frontier have drawn us upward, forward, on to the next milestone. This book addresses the question of what the next hundred years will bring in space. Two big unanswered questions serve to frame the future: Can we live in space inexpensively, mining water and other resources needed for human life from celestial bodies? Is there any commercial payoff from the settlement of the inner solar system? The answers to these questions will determine whether or not space settlement is our manifest destiny.

This look toward the future frames a painting which illustrates a number of important space-related issues. What sort of exploratory ventures are human beings willing to undertake? Do we have the guts to make those bold steps that visionaries are urging upon us—and does it make sense to do so? What roles will humans play in space in the next century? What is the potential for space commerce? How will the military be using space, and what will be the impact of the military use of space on civilian activities?

The year 2000 is nearly upon us; it's tempting to look ahead to the more distant future in considering what may happen. In fact, it's not just tempting, it's necessary, for what we do in the near future will set the course of the space program for the next century. If we are farsighted, we can lay the groundwork for future expansion into the solar system now. As we do so, we will be able to determine whether such an expansion makes sense or not. With lack of foresight, we will embark on yet another megaproject which will last a decade and then leave space enthusiasts

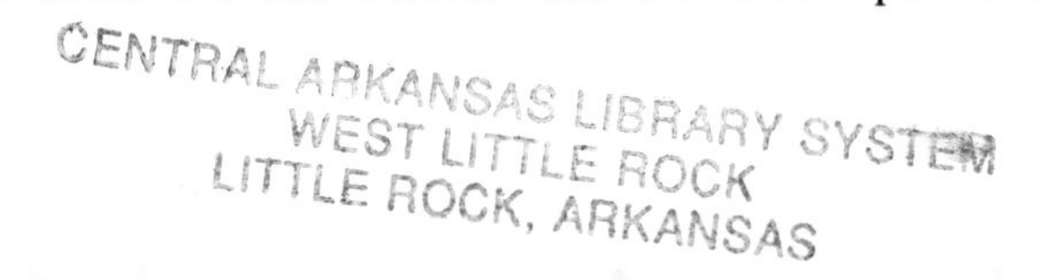

wondering what to do for an encore, as was the case when we landed on the moon.

This book is bolder than my earlier book (*Space 2000: Meeting the Challenge of a New Era*), since I'm relying less on what we're doing now, but I do try to keep my feet on the ground. It's not all forecasts of the next hundred years; I'm using the visionary question as a way of focusing the discussion of some issues like the human role in space, asteroid mining, space commercialization, space militarization, and the future of space science.

I've been working with the space program in one way or another since 1974, as a guest investigator on various satellites operated by NASA or its European counterpart ESA, on various review panels, and as a NASA grantee. However, I've never been a NASA employee, nor have I received a security clearance; the views I express below are my own and are based on the open literature.

I thank many colleagues, including Peter Banks of Stanford, Dick Henry of Johns Hopkins, Carol Hoffecker of the University of Delaware History Department, Norman Ness of the Bartol Research Institute, Dana Rotegard of Rotegard Venture Capital, and Cheryl Thompson of the University of Delaware Library for help in providing me with useful comments, materials, and stimulating ideas. Linda Regan and Victoria Cherney of Plenum have provided the feedback and prodding that all good editors should. My children, Alice and Tom, and my wife Wendy have tolerated the weekend and evening hours that a book requires; it is only now that Alice has an answer to her question: "When will we go bowling again, Daddy?"

Contents

PART I

The Shores of Space

The exploration of space has just begun. Many space enthusiasts see the journey into space as the next logical step in a human venture of exploration and discovery which began 500 years ago in the time of Christopher Columbus. However, the differences between what we're doing now and what Columbus did in days gone by mean that few people who are not already space enthusiasts find that the memory of Columbus is, in itself, a convincing reason to go into space. Now that we've passed many of the early milestones in space exploration, having launched satellites and human beings into orbit and landed on the surface of the moon, a number of alternative space futures open up. What we do depends on whether space exploration is purely government-funded science or whether it is commercial, and whether keeping human beings alive in space will always be as expensive as it is now.

Another factor governing what we will do in space is whether we have the political courage to support a long-term, sustained exploration program. One way of determining just what kind of an exploration program human societies will support is to examine the past. While there is an exploring spirit, a desire to go where no human has ever gone before, this spirit by itself has never been the sole reason that humans have ventured forth into unknown oceans. The past can serve as a guide, suggesting just what kinds of space programs are politically possible. The immediate past, the first 30 years of space exploration, have also shaped the major space agencies and set the scene for our future in space.

What comes as a result of all this is a number of possible space futures. We may indeed settle the solar system, first establishing outposts and colonies in low earth orbit, and then gradually expanding the human sphere of influence to more distant objects like Mars, its satellites, and possibly back to the earth's Moon. However, if history is any guide, space settlement must pay off in practical terms if it is to occur. It may be too expensive, too pointless, or both to send people into space; in that case, we may just launch robot mines, factories, and laboratories into space, with people playing a more secondary role. While sending humans to Mars is probably part of any one of these space futures, the nature of the expedition—whether it is a precursor to outposts, colonies, and settlements or whether it is a one-time symbolic journey—will depend on the larger future of the space program.

CHAPTER 1

The Next Frontier?

Where are we going in space? Forget, for the moment, short-term concerns about a thriving Soviet space program and a crippled NASA; try to take the long view and look a century ahead. Even now, computer-controlled machines have hurtled billions of miles from the earth, providing us with a brand new perspective on our solar system and on our universe. Humans have landed on the moon, and have survived in space stations for nearly a year. We've come a long way since the early days when astronauts were strapped into oversized metal cans, tossed into orbit, whizzed around the earth for a few hours, and quickly came back to the comforts of home.

And yet much may remain to be done. Astronauts are no longer helpless test objects who can barely crawl out of their space capsules by themselves. Yet they are still only short-term visitors to space, not permanent residents. The human presence in space is confined to low earth orbit, only a few hundred miles up from the earth's surface. Space settlement or colonization, although much discussed, is still a dream. Will this dream become reality?

Transforming that dream into reality, if it is to happen, will take time. Even the automated exploration of space has only begun. Orbiters and remotely operated landers have landed on Venus, the moon, and Mars, but these landers weren't mobile, and could only provide intensive exploration of one place. Telescopes in space, above the atmospheric murk, have been small ones, with useful lifetimes of a decade or less. More ambitious facilities are probably considerably closer to becoming reality than space colonies are, but space scientists' appetites for greatly expanded facilities far exceed NASA's financial ability to provide them.

Humanity has been in this position, looking out onto unexplored emptiness, before. About 500 years ago, one of the most dramatic moments in the Western European exploration of the earth came when Vasco Nunez de Balboa first sighted the Pacific Ocean. Balboa, enmeshed in a political struggle over the control of the Spanish settlement at the Caribbean island of Santa Maria del Antigua, sought to do something important to save his position as governor. Native Americans had told Balboa that there was a great sea to the west, where the natives had lots of gold. Balboa's expedition of several hundred people took three weeks to hack through jungles, wade and swim through swamps, and climb over the thousand-foot mountains which split the 45-mile-wide Isthmus of Panama. On September 25, 1513, Balboa commanded his army to halt and went alone to the top of a mountain peak. The vast ocean shimmered in the sunset, looking deceptively peaceful. So Balboa named it the Pacific, unaware of the treacherous storms which would buffet fleets of ships sailing around Cape Horn in the centuries to come. He could scarcely have realized how important this ocean would be to Western Europeans and Americans, and how pivotal his discovery was.[1]

In many ways, we are now in the same situation that Balboa was in five centuries ago. Balboa and his contemporaries had heard tales of gold to the south; now similar visions of space colonies and potential industries are the dreams which draw some people, at least, out into the depths of the Universe, or at least as far as Mars, hundreds of millions of miles away. Visionaries of Balboa's time speculated about the tremendous profits which could come from additional voyages, in the same way that space enthusiasts now write about lunar bases and trips to Mars.

There are, of course, important differences between our vantage point and Balboa's. We have known for centuries that outer space exists, and so the Soviet's first satellite Sputnik did not "discover" a previously unknown entity. The discoveries of the past 30 years have opened up space only in the sense of showing that, in principle, outer space can be explored by humans. It's much more difficult to send people out into outer space than it is to build a boat and sail on the Pacific Ocean, but it can still be done. We also have the ability to send automated probes into the solar system to extend the range of the human senses, making it possible to explore regions of space without actually sending astronauts there.

Many of us have attempted to look ahead to the near future, to the next 30 years or so of space exploration. Recent books like my own,[2] those of others,[3] and NASA internal planning documents can forecast the near future by describing a number of projects like the space station which are currently in process and are logical extensions of what's been done before. A more distant vision of the future, of the next hundred years, is a bit more difficult to come by. However, such a vision can help shape the decisions about what we do in the immediate future. Because space exploration is so much more difficult than the exploration of the earth, a successful human leap off of the confines of the earth's surface probably can't afford a great many false starts and failed initiatives. We've just suffered through one such false start with the space shuttle, which was not the space truck that it was advertised to be.

Many authors who write about the space program, including the members of the National Commission on Space, seem to think that space colonization is inevitable. In their view, space must be settled in the 21st century in the same way that Europeans occupied and settled America a few centuries ago. The phrase "manifest destiny," drawn from American history, recurs again and again. Is this really so?

Realistically, though, space settlement is not inevitable. Whether we human beings settle the inner solar system depends on the answers to three critical questions. Although we can't definitely answer any of them, they serve as a framework to consider various possible future trajectories of human exploration of space. These questions are[4]:

- Do we, as a nation, have the courage and vision to make the effort needed to explore and, if possible, settle outer space?
- Can the air and water needed to support human life be found or extracted from extraterrestrial bodies, or do these have to be transported into space from earth?
- Will space commerce develop beyond the communications satellite business, and what is the nature of commercial opportunities in space?

These three questions will frame this book. I will address them initially in this chapter, and more deeply later on. Space enthusiasts assume that the answer to all three questions is an unequivocal, resounding "yes." If these positive answers pan out, they lead to optimistic pictures of lunar bases, space settlements, outposts on Mars, and other ideas which would have seemed like science fiction 50 years ago. However, positive answers to these questions are not inevitable, and alternative space futures emerge as equally valid prognostications.

THE FRAMEWORK OF THE FUTURE

Our Political Will

Space travel is a very special and difficult human venture. The space environment is very hostile. There is no air or water to sustain human life; all of the supplies which must sustain our astronauts and our machines have to be brought up from earth. People and electronics have to function in an environment which is quite different from the centrally heated or air-conditioned comfort which we are used to. The establishment and continuation of a space exploration program requires a tremendous investment of courage, energy, and resources. We may have the technological ability to establish space stations, outposts on the moon, and send human beings to Mars, but do we have the courage to implement our vision? The costs of space travel are so high that a decision to go to Mars or wherever is inherently political, whether it be in the United States or the Soviet Union. Spark Matsunaga used the term "political will" to refer to the collective courage of a democratic society to implement a vision.[5] This term refers to the confluence of forces which lie behind any national effort; this political will is really a combination of political, economic, and social factors. The words seem equally appropriate to a bureaucratically organized state like the Soviet Union, since the necessary commitment to something like a Mars trip or space colonization will require a commitment of a whole political system rather than one leader who may only be in power for a decade or less.

The futuristic perspective of this book makes it imperative, however difficult it is, to look beyond short-term political concerns. The late 1980s see the American space program stumbling in the dust, buffeted by the loss of the space shuttle *Challenger*. Few in the White House or in Congress seem to care; the necessary money to restore America's preeminence, if it were possible to do so, is simply not there. The stock market collapse in October 1987 crystallized a previously existing preoccupation with the budget deficit. Such a political climate is by far the most difficult one in which to sell a new space program. Even were the budgetary situation more relaxed, NASA's aging top management seems unable to provide the spark necessary to translate bold new visions into reality.

If past history is any guide, this difficult situation will not persist for the next hundred years. Unless America is really going down the tubes, there will be times in the next decades when young, vigorous leadership is not confronted by seemingly uncontrollable Federal deficits and a hostile or indifferent Congress. Furthermore, even if America doesn't seize whatever opportunity is available to venture forth into the universe, other countries can pick up the ball which we may still let fall to the earth. In the late 1980s, the Soviet Union reveled in the successes of their thriving space program in a series of publicity triumphs connected with the 30th anniversary of the Sputnik launching. The European Space Agency and the Japanese are not that far behind the two superpowers.

How does one transcend any temporary political situation, whether it be especially good or especially bad, and realistically ask what sort of space program could be sustained over a long time period? I believe that only history can provide an answer. Because the space program must be led by nations rather than individuals, it is the collective political will that will sustain or fail to sustain a particular kind of space program. As a result, I contend that the best way to ask what kind of space exploration program is sustainable over the long term is to look to the past and ask what kinds of exploratory ventures have succeeded then.

The space program can in some sense be seen as the continuation of the exploratory efforts of centuries. From a Western European perspective, these efforts began in the modern era with Columbus's discovery of America and continued, with various important changes of emphasis, through the exploration of Antarctica which goes on today. The exploring spirit, the need to climb a mountain just because it's there, is the driving force behind such ventures. There is no shortage of people who have the exploring spirit.

A space program, however, requires more than just courageous astronauts. It requires a political system—in Western democracies, taxpayers—to support them, to pay for the spacecraft, engineers, rocket fuel, and other support personnel which make it possible to send people to the moon and beyond. The historical, political, and social context of the space program become important because there are significant differences between the exploration of space and the exploration of the earth. These differences need to be remembered by naive idealists who think that one need only mention the name of Columbus to convince everyone that we need to go to Mars right away. A historical consideration of these changes can provide some perspectives on what sorts of space programs can be considered to be realistically sustainable over the long pull.

What sorts of governmentally supported exploratory programs are sustainable over a century? A look back at the recent and distant past provides a clear indication that Apollo-type moon landing programs, focused on clearly identifiable milestones which are goals for their own sake, cannot lead to a long-term program of exploration. The sustained exploratory programs have provided some kind of tangible, commercial return, either a short-term one to the explorers themselves or a long-term one to a government eager to protect its present and future interests. Knowledge and adventure, no matter how admirable for their own sakes, can drive individual explorers to attempt all sorts of improbable feats but have not generated the political will needed to make these programs self-sustaining.*

Human Use of Extraterrestrial Materials for Survival

The second of the three big questions outlined above, our ability to utilize space resources in order to keep space settlements going, is one of the ways in which nature will determine both what we can do and what it makes sense to do. This is really one side of a cost/benefit approach to the space program. How much will it cost us to operate in space, to perform materials science experiments, to mine asteroids, to observe the near and distant universe, and to fix communications satellites? While it might seem too narrow to focus on the use of extraterrestrial materials, it is the major cost item for space settlement. Is it necessary to lift life-supporting materials from the earth at great cost or can humans in space make use of materials which are available locally?

The futurist Buckminster Fuller coined the phrase "spaceship earth" in the 1960s in an effort to sensitize people to the closed nature of the earth's ecosystem. The phrase was intended to evoke memories of the earth seen from space, and used to emphasize that nothing that we use on earth is really lost when we "throw it away" or, more realistically, put it somewhere where we can then forget about it. Since I'm basically a technological optimist, I wish I could say that this phrase really applied to real spaceships. Unfortunately, contemporary spaceships throw a great deal of stuff away, and the need to do so increases the cost of human space flight immeasurably.[6]

Currently, humans living in a spacecraft consume tens of kilograms of material per day in order to stay alive. Most of this stuff is water—drinking water, wash water, water to flush the toilet with, and water to cool and operate the spacecraft. While some limited reuse of water is possible in principle, in practice most of this water is brought up from earth at great expense, used once, and then dumped over the side, forming

*See Chapters 2–4 below for a more comprehensive discussion.

a little cloud around the spacecraft. Currently it costs about $5000–$10000 to hurl a pound of material from the earth's surface to orbit.[7] With these launch costs, you need to spend a few hundred thousand dollars per day to keep a person in space supplied with water. It's very expensive to live in space, if we support astronauts in the way that the major space powers do it now.

No matter what happens to other aspects of space-flight economics, it is clearly impossible to imagine large numbers of humans existing in space if it costs hundreds of thousands of dollars per day per person to maintain them. Part of the problem is that current spacecraft technology permits only a limited amount of recycling of water, and no reuse of other materials. Even if new technology could cut the water requirements tenfold or even more and even if launch costs drop tenfold as well, the cost is still very large. You need to cut the cost by 10,000-fold over the present value to cut the cost of living in space to the cost of an expensive hotel room. It isn't just water, either; requirements for food and air are significant as well.

There is a way around the high-cost dilemma. Space is not empty—it contains large bodies like the moon and Mars, and smaller objects like comets and asteroids. If air and water can be extracted or captured from any of these objects and used by human beings, the resource requirements for space colonization would plummet precipitously and become much more reasonable. Such a colony could survive in much the same way that early colonies in America survived, with supplies from the mother country consisting of small transfusions of genuinely irreplaceable goods rather than huge, bulky, expensive cargoes.

We don't know, now, whether the use of extraterrestrial materials is a pipe dream or is something that can become real, and so both options have to be considered in any discussion of the future of the space program. Part of our ignorance derives from having only just begun to explore the solar system. For example, we know that Mars has at least one polar cap which contains water, but we don't know how accessible that water is. Another part of the dilemma is technological. Little effort has been made by either Americans or Russians to recycle materials in space stations, for various reasons, and so we don't know whether the material requirements for sustaining life in space need to be as extravagant as they are now.

For the moment, consider that there may be two different answers to the question of possible use of extraterrestrial materials. The materials we need—principally water, but including oxygen and other substances—may be abundant on the moon, on Mars, or in asteroids and comets which come near enough to the earth so that they can be captured. Alternatively, these materials may remain very inaccessible, in lunar rocks that are hundreds of kilometers beneath the moon's surface, or trapped beneath impenetrable crusts on Mars or hidden in comets. Both of these possibilities suggest different possible futures for the space program.

Space Industrialization

The cost of settling and using space is only one side of the usual cost/benefit consideration which enters any human decision. What are the benefits? Increased knowledge and a sense of adventure are there, to be sure, but the history of human exploration of the earth demonstrates that exploratory ventures which have some potential for commercial return are likely to be far more vigorously pursued than ventures which are undertaken for the sake of science alone. Even the most enthusiastic proponents of space colonization agree that the space colonies need to provide some economic return for the money which is invested in them.

The term "space industrialization" has been used to describe an enormous variety of hypothetical activities. I consider an activity in space to fall within the purview of "space industrialization" only if a space station, structure, or colony provides some products or services which are ultimately sold to customers outside the aerospace sectors of industry or the Federal government. Furthermore, a successful industry, in normal terms, should generate substantial rather than incidental revenues outside the aerospace and government sector. Most emphatically, I don't consider space to be "industrialized" just because some company has built a space station or platform with its own resources and then leased this facility to some government, since in such a case the government is the ultimate source of the revenue.

An excellent example of space industrialization is the communications satellite industry. While the satellites themselves are sometimes launched by private corporations like AT&T and sometimes launched by governments, the revenues are generated from billions of customers, ranging from individuals making telephone calls to banks using satellite communications to maintain international networks of automatic teller machines. One of the reasons that we are in space at all is that Arthur C. Clarke, the noted science fiction writer and space pioneer, recognized the potential of satellite communications as early as 1945 and promoted this concept actively in the 1950s.

Why is space industrialization important? With it, there are a whole host of private interests which can support further space exploration. Without it, the future of space exploration depends on some government or other—be it American, Russian, Japanese, or whatever. The past history of the space program and of other exploratory expeditions shows that once some milestone is reached, once the first astronaut lands on the moon for example, public interest and support plummet. While some modest level of purely scientific, noncommercial activity in space could probably be sustained on the basis of national prestige and knowledge gathering, such a program would undoubtedly be much more modest in scope than a commercially based space program.

Forecasting the potential commercial development of space, especially over 100-year time horizon, is difficult indeed, and I won't presume to try to do it. There has been much visionary talk about such activities as solar power satellites, materials processing in space, and asteroid mining. But despite many efforts, nothing else like the communications satellite industry has made the crucial transition from dream to reality in 30 years; fledgling space industries like earth photography and materials processing are not yet genuine commercial successes. When I look ahead to the 21st and 22nd centuries, the best I can do is to leave this question open, discussing the possibilities and staying aware of the need to distinguish between the visions of salespeople and what's likely to become real.

The Space Program in Context

Perhaps it's natural for the reader to presume, as a matter of course, that what follows will be true in the 21st century, but I believe it's worth setting forth some of the assumptions I'm making about the 21st-century world when I make predictions about our future in space. A space program will not happen if the world is undergoing economic or political chaos. Thus I'm assuming that some solution to the immediately perceived problems will be found and that the 21st-century world will be a relatively peaceful, not a chaotic, one.

No one knows who the major economic powers will be, but it doesn't really matter since all of the major candidates for economic dominance in the 21st century are reasonably active in space.[8] All of the major powers on historian Paul Kennedy's list of the major players in the late 20th century—China, Japan, Europe (considered a collective nation), the Soviet Union, and the United States—have substantial space programs, with China's the weakest. Evidently, all major nations have decided that having some kind of space program is an important part of being a major power. The signficant commercial potential of space is, perhaps, the driving factor.

Like Paul Kennedy, I am presuming that there will be no major unpredictable environmental or social factors which will profoundly disturb the 21st-century world. Some will find such a presumption foolish; one fellow scientist, finding out what book I was working on, asked me what we would use for energy in the 21st century. His clear implication was that we would all be freezing in the dark and that automobiles—not to mention interplanetary rockets—would be historical relics. It's quite possible that such apocalyptic visions of the future may come to pass, though the past is loaded with predictions of future disaster that never happened. The predictions in the 1960s of global famine in the 1980s were as far off the mark as the predictions that the New York City telephone system would suffer from terminal strangulation by the 1950s. In any event, apocalyptic visions of the future are for another book—not this one. My presumption is that at least one of at least one of the major space powers—America, the Soviet Union, Europe, or Japan—will remain economically healthy through the 21st century. Not that it matters, but my vision is that we will be using solar energy 100 years from now.

FOUR VISIONS OF THE FUTURE

I have posed a number of questions which will determine the future of our space activities. The question of our political will—the extent to which we will be willing to support a particular type of space program—will determine how far we will go in any particular direction. It's useful in prognosticating to separate what is technically possible from what is politically feasible, and so the two technical questions will be used to frame the possible visions of the future:

- Can life support materials be obtained from celestial bodies, or must they be launched at great expense from the earth's surface?
- Will space industrialization happen?

How does asking these unanswerable questions help make progress? With two questions, and two answers for each, we then have four possibilities or scenarios, set forth in Table 1 and described briefly in the rest of this chapter. My reason for focusing on these questions is that we can answer them in the next decade or two, with the right sort of space

Table 1. Space Futures

		Will Space Industrialization Work?	
		Yes	No
Can extraterrestrial resources be used to support humans in space?	Yes	Full space settlement	Research and tourism
	No	Robot mines, factories, and labs	Space science only

programs. In addition, these four possibilities or scenarios cover the ground of reasonable futures.

The four scenarios, set forth briefly in the table, serve to delineate the possibilities. Each entry in the table corresponds to one set of answers to each of the two purely technical big questions. At the upper left is the "space settlement scenario," the dream of space enthusiasts. If extraterrestrial materials can be used to support life, and if space commerce becomes a reality, then the answer to both of these big questions is yes. Then, and only then, it's likely that human beings will settle at least some parts of the inner solar system, in the "full space settlement" scenario.

Other table entries correspond to other possibilities. If life support continues to require large transfusions of material goods from the earth, then the human role in space will be far more limited, as indicated in the second line of the table. In such a case, space industrialization could still happen, leaving us with a case which I call "robot mines, factories, or labs." Another possibility is that space industrialization also wouldn't happen, leaving space to the space scientists in the "space science only" scenario. In my view, the least likely possibility is that shown in the upper right of the table, where space resources are abundant but there's no commercial incentive to settle in space. I call it "research and tourism" because space tourism is the only human activity, other than research, which I can conceive of in such a scenario.

Full Space Settlement

This is the scenario described in many visions of our future in space, and is clearly the one that many space enthusiasts dream of.[9] The ingredients of this scenario include such elements as large space stations with staffs of dozens, permanently inhabited lunar bases, human outposts on Mars, cycling spaceships making the run to Mars and back in the same way that jet airplanes cycle from New York to Paris and back, and space colonies. Not all of these elements need be present, of course, but most space enthusiasts include many of them in presenting optimistic visions of the future.

One "full space settlement" scenario might run something like the following[10]: American and Soviet space stations, operational by the turn of the century, evolve over the following decade or two into extensive, flexible space laboratories, factories, and spaceports. By the third decade of the 21st century, product sales and fees from users of laboratories in space become large enough that much space activity no longer depends on any federal government, be it American, Russian, or Japanese. During this time, robot probes have been exploring the surface of Mars and its satellites, finding layers of icy permafrost—permanently frozen ground—only a few feet below the surface. Astronauts living in the space station for periods of a year or more can establish just what needs to be brought along on the Mars trip in order to keep humans alive and thriving during long stays in space. The first human expedition to Mars, in 2035, leaves more than just the American, Soviet, Japanese, and newly designed "European" flags fluttering in the thin Martian breeze. Space hardware used to propel humans to Mars remains on the Martian surface and in orbit, to serve as the nucleus of a Martian base and spaceport. The spaceport is complete in 2050, and by 2080 cycling spaceships make the Mars run every two years or so, bringing a new crew and the small nucleus of vital supplies which cannot be mined from the Martian soil. The highly touted human expedition to Mars would simply be a part of humanity's slow expansion toward the red planet and its satellites. Up to 30 people could be accommodated in such a base, with individuals living and working in the vicinity of Mars for periods of years.

In the latter part of the 21st century, these early bases start to evolve into colonies. First, in 2050, comes an early settlement in orbit around the earth, which is increasingly supplied not from the earth but from Martian material and from asteroid chunks which are captured, nudged into orbit near the space station, and used as sources of water and ore. By 2070 the first child is born in Space Colony One, conceived and reared by a couple who have spent nearly a decade of their lives in space. The first citizen of Phobos, the Martian moonlet, sharing the Soviet and American nationality of her parents, cries to meet the dawn on New Year's Day 2100.

Such a story line is both amusing and inspiring; the dates even make it look real (though, of course, it's only imaginary). Are the last few paragraphs at all reasonable, or are they destined to remain fantasy forever? I buried in my description of this scenario two vital ingredients which may or may not exist: readily accessible sources of life support materials, principally water, and the possibility of genuine commercial activity, space commerce in the sense that I identified it above. If one of these key foundations for the space settlement scenario does not come to pass, then our activity in space will be considerably more limited.

This "full settlement" scenario seems to be the one that many space planners and visionaries regard as being inevitable. A number of vocal space scientists, who see no evidence that the two key ingredients are at hand, see this scenario as impossible. In particular, a contrasting view in the space community has been provided by space scientist James Van Allen, who has called for the complete abandonment of the manned space program.[11] One source of the communications gap between space enthusiasts and space scientists is probably that space enthusiasts regard this scenario as inevitable and that some vocal space scientists are sure it's impossible. Both extreme viewpoints may be wrong. Suppose that one of the two foundations of the space settlement scenario is missing. A number of other possibilities then must be considered.

Robot Mines, Factories, and Labs

Suppose that commercial activity in space proves to be possible, but that the expense of maintaining human beings in space continues to be considerable, in the million dollars per day per astronaut range that it's in now. Are humans necessary for commercial activity? The visions of many space dreamers were shaped, in part, by the science fiction literature of the 1940s and 1950s, and since then the capabilities of computers and of machines they control (in other words, robots) have expanded enormously. Perhaps humans are not necessary at all, or are only needed in order to occasionally visit remotely operated space mines and factories, which could either function autonomously or could obey radio commands from the earth.

In this scenario, space would be "settled" by robots! Scores of space factories or information handling facilities would orbit the earth. Remotely operated platforms would supply power, communications, and docking facilities. Depending on the raw material that these factories need, they might be supplied by robot mines on the moon, or by fetching asteroids from the supply of earth-crossing ones. At last resort, some materials could be brought up from earth, but at great cost.

Machines aren't perfect. As a result, there would be an important but limited role for astronauts in this vision of the future. There could certainly be orbiting repair facilities where astronauts could repair or replace defective components. Depending on the machines' reliability, there might be a few repair people living in a permanent space station, or more likely, astronauts could go into orbit to fix something that wasn't working and that was more easily fixed than replaced. Astronauts would probably also be present in some kind of space station or platform in order to work with prototypes of new facilities.

Research and Tourism

Another scenario is the mirror image of the robot mines and factories one: The ingredients for life support are available somewhere in space, but the opportunities for genuine space commerce remain limited to the communications satellite business. Even though it would be cheaper to sustain life in space than it is now, large space colonies involving hundreds or thousands of people, and settlements on distant, hard-to-reach planets like the moon and Mars, would seem difficult to justify. What would those people do?

Some research scientists could take advantage of space tourist facilities as ways of supporting their existence in space. A lot of space science can be done by remotely operated instruments, but some science does seem to require astronauts in orbit. Studies of the human reaction to low gravity, made for their own sake rather than for any possible application to future space flight, might require the presence of astronauts. Materials scientists might be needed in order to study the process of crystal-growing in space.

Who else might travel into space? Tourists, perhaps; Society Expeditions of Seattle, Washington, has collected deposits on 265 reservations for a 12-hour sightseeing tour. Looking toward the more distant future, very-well-heeled fat cats might be willing to pay a million dollars or more for an absolutely unique vacation. Perhaps an enterprising travel agency could (for a hefty price) attach another power module and another living unit to that space station, sharing docking and communications facilities with the space station.

My own guess is that this scenario is least likely to be realized in the future, largely because I find it hard to believe that a few scientists and tourists could stimulate the capital investment (for instance, in water recovery plants on Mars) which could cut the costs of living in space to fit a tourist's budget. Indeed, a recent report by the Space Science Board of the National Academy of Sciences mentioned the possibility that no science, other than the life sciences, could justify the human presence in space.[12] My primary reason for introducing this idea is to highlight the severe requirements of space settlement. Recall, then, that the framework of this whole discussion is not to predict the future; rather it is to outline ideas which, in some unknown combination, will probably illustrate where we are going in space.

Space Science Only

If supporting human beings in space remains expensive, and if commercial use of space remains a dream, then what's left to do up there? Space science doesn't require the presence of humans in space, and some of NASA's most spectacular achievements have been in the area of using space to extend the frontiers of human knowledge. In most areas of space science, experiments and observations made so far have merely scratched the surface and left a plethora of unanswered questions.

For example, consider our understanding of our own planet. The earth is a complex, interrelated system in which the oceans, surface, and atmosphere interact to produce ocean currents and weather. For centuries, earth scientists have faced the difficulty of making observations and measurements which cover the entire globe rather than a small, readily accessible piece of it. It's easy to maintain weather stations in populated areas, and not too difficult to obtain reasonable coverage of well-traveled shipping routes. But only space can provide the necessary information about remote areas like the Arctic, the Antarctic, and many remote oceans. Earth scientists have developed many clever ways of measuring important quantities remotely, from orbiting satellites. Twenty-first century earth scientists could use these techniques to, for example, predict hurricane and storm tracks quite precisely.

We can look up, out into the distant universe, instead of looking down. The earth's atmosphere blocks most types of electromagnetic radition—X rays, ultraviolet, and infrared radiation—from reaching the earth's surface, providing astronomers with a very narrow window onto the universe. Observations from space have unveiled a new, more violent universe, in which titanic explosions can rip stars and galaxies apart. Work so far has shown us that the universe is violent, but space telescopes have been small and limited in capability. The next hundred years of space astronomy offer the possibility of real understanding of cosmic violence and, indeed, of our origins.

Many other areas of space offer similarly enticing vistas. The solar system provides a natural laboratory of planetary-sized bodies which are similar to, yet significantly different from, our own earth. Why is Venus so much hotter than earth, even though it is virtually the same size and is only somewhat closer to the sun? Why is Mars so cold? Answers to these questions can help us understand our own planet better. The space environment, in which the force of gravity is almost entirely absent, is a new laboratory for studying the behavior of matter and the behavior of living systems. All areas of space science are currently very active, and could continue to remain so for a century or more if the resources are provided.

DESTINATION: MARS AND BEYOND

Mars

Will we go to Mars? Of course we will, someday. I suspect that no matter what happens in the 21st century, a human trip to Mars will be part of it. I find it inconceivable that we would be technologically able to go to Mars and not decide to go. A trip to Mars could be done now, with current technology, though it might be rather more expensive than it could be if we can design the journey rather cleverly with the help of precursor journeys with automated probes. However, the nature of our Martian journey can vary considerably.

One way to get to Mars is to indulge in some kind of crash program, like the Apollo mission to the moon. During the Apollo program, little thought was given to the development of some kind of space infrastructure which would leave a heritage in the form of useful hardware. Apollo was a symbolic journey, an episode in the Cold War between the United States and the Soviet Union, as well as our first step in bringing humanity into space. The Apollo program left little space hardware that could be reused for a long-term program, and that which we could have reused we threw away.

Space enthusiasts, people who believe that the space settlement scenario is likely or even inevitable, are rightfully cautious about a sudden Apollo-like push toward the red sands of Mars. Think about what would happen after the successful completion of such a mission. Then we would have landed on Mars, and could no longer use the journey as the carrot which would draw public support for long-term investment in space infrastructure. If we are to establish self-sustaining colonies in the inner solar system, space enthusiasts must persuade the taxpaying public to make a relatively large investment in space hardware and infrastructure which will have a payoff, if any, lying far beyond the next election.

Even if space settlement doesn't happen in the next century, I'm convinced we will go to Mars anyhow. It seems clear that the Soviet Union is definitely thinking about a trip to Mars in the long term. Will American political leaders stand still and let the hammer and sickle flag flutter alone in the thin Martian breeze? I doubt it.

Beyond Mars

Projecting our space activities beyond a Martian outpost becomes risky indeed if you worry about whether or not words written now have any relationship to reality. Some readers may think that in recounting some of these more distant possibilities I may have stepped over the boundary separating science from science fiction. Yet these possibilities are fun to think about, as long as you realize that they are speculative.

When (and if) we have established a human presence somewhere in the inner solar system, further expansion through the solar system becomes considerably easier. There are hundreds of thousands of claimable chunks of real estate in the asteroid belt, and some of those objects may contain the liquid of life—water. Personally, I find it relatively easy to imagine space settlement expanding from a colony near earth to many colonies in the solar system, and much harder to visualize how the colony near the earth may come about in the first place. As Princeton physicist Freeman Dyson puts it, our descendants may, someday, be "homesteading the asteroids" in the same way that 19th-century Americans pulled up stakes in the East and claimed their 160 acres in the West.[13]

Even more speculative is a scenario—called "terraforming" by its inventor Carl Sagan—in which entire planetary environments are modified by human activities in order to make them more habitable. Is Mars too cold? Melt the polar caps, make the atmosphere thicker, and warm it up. Water-formed channels on the surface of Mars tell us that at one time in the past, Mars was much warmer than it is now, possibly warm enough to sustain human life. Can human ingenuity undo what nature did to toss Mars into the deep freeze?

Perhaps the most far-out scenario is one which has become quite familiar as a result of the *Star Trek* television series—interstellar space travel. Most science fiction writers have to assume that something like "warp drive" or "hyperspace" exists in order to have a story, and such schemes for travel faster than light have been allowed for years as legitimate in the science fiction literature. As a scientist, I have to regard breaking the light barrier as being out of bounds, within the context of present-day science.

While these ideas are highly speculative, they still lie within the boundaries of modern science in that no scientific (as opposed to engineering) advances are required in order to see them happen. None of these ideas has developed to the point where it makes sense to plan for them to happen, but they are certainly fun and mind-stretching to think about.

* * *

Two basic questions frame our future activities in space, taking the long view into the 21st and 22nd century: Can extraterrestrial materials be used to support human life in space? Will space commerce exist? Based on our current knowledge, the correct answer to either question could be yes or no, and the combination of possibilities leads to four scenarios for the future of space travel.

But there is yet a third question which lies in the background of all this: Do we have the political will to make the substantial investment of human and financial capital in order to realize any of these possibilities? I don't want to pin any hopes on a hypothetical change in human nature, so it seems that the best way to answer this third critical question is to look toward the past, which is considered in the next two chapters.

CHAPTER 2

The Exploring Spirit

Many space enthusiasts feel that there is no need to really answer the question of why we are going into space in the first place. "Because it's there" is seen to be grounds enough. Konstantin Tsiolkovsky, a 19th-century Russian high school teacher who was the first to describe the scientific principles of rocketry as applied to space flight, penned a frequently quoted sentence: "Earth is the cradle of humanity, but one cannot live in the cradle forever."

Words like these are invocations of some kind of existential "urge to explore," an internal human drive similar to, but in most cases less important than, the primal human desires for such things as food and sex. Following historian Daniel Boorstin, I'll call this the *exploring spirit*.[2] We all know it; it's what made me, as a kid, skate down the shallow Park River in back of my house, seeking to discover what lay behind the small groves of trees which grew in our, and our neighbors', backyards. (Incidentally, the Park River was shallow enough so that even with the conservative retrospective of adulthood, these voyages were quite safe; when I fell through the ice, I only got my feet wet.) Most human beings have such urges. Americans, in particular, value exploration in and of itself because of the importance of the frontier in our history. People interested in the space program are probably self-selected as being those whose internal makeup contains a substantial amount of the exploring spirit.

What is this exploring spirit? It's what's left behind the motivations of Columbus and his contemporaries when you take away the commercial and religious motives of his financial backers. It's the common element found in all the voyages of discovery. It's a combination of pride, curiosity, and a heroic sense of adventure.

THE MAJESTY OF SPACE

The illustration on the dust jacket of this book, reproduced below in black and white, represents one way of capturing this exploring spirit, particularly as it relates to the space program. The alien character of the space suit cannot hide its most precious contents—a human being, protected by this technological device from the horrid vacuum of outer space. The jet-propelled backpack allows astronaut Bruce McCandless to float truly freely, not tied to his mother ship by an umbilical cord, a celestial apron string which might prevent McCandless from feeling like a space inhabitant and not just an extension of the space shuttle. With 20th-century technology, we can suit people up, push them out the airlock, and let them move around and do work in space. Pictures like this evoke the exploratory urge among many of us.

Can it possibly be that this alien universe, most of which consists of an airless, cold vacuum, can become part of the human territory? The human species, unlike any other, has extended its geographic range throughout the earth. Is it our destiny to go still farther? Of course it is; if we can send people to the sands of Mars, we will, even if it's only to walk once on the Martian surface.

To some, photographs of astronauts freely floating in the vacuum of space, of people setting forth on the moon, of the windswept, reddish sandy plains of Mars provide reason enough to go into space. But to others, they seem almost irrelevant. Skeptics often ask, "Don't we have enough problems at home?" One response is that there is an exploring spirit which is an inexorable part of human nature. But what is it? What are its limitations? Can it, realistically, be used by itself to sell a space program in the political arena? In this chapter, I'll examine the exploring spirit by looking at the powerful human feelings which go into it.

PRIDE

Pride stirs deep emotions, both in the explorer who makes the journey and in his countrymen and fellow human beings who share in the achievement. For the explorer, it's a matter of ego, of being the first human being to set foot on a particular place. Columbus demonstrated his need for ego gratification in his demand to be known as the "Admiral of the Ocean Sea" as a result of his discovery of America. Not all explorers are so self-centered; Apollo 11 astronauts Armstrong and Aldrin "came in

Astronaut Bruce McCandless floats in space, with the jet-powered backpack his way back to the space shuttle. (NASA)

all mankind" and didn't insist on being called admirals of Nevertheless, when you read the Apollo 11 astronauts' mem- they were all overwhelmed and happy that the way in which astronauts rotated through the Apollo flights happened to result in the three of them being assigned to the Apollo 11 flight, which turned out to be the historic lunar landing.

Throughout human history, though, the explorers themselves have been a minority of people who needed to rely on the efforts and support of others in order to venture forth. The Apollo 11 astronauts landed on the moon in 1969, a difficult period in American history, when the nation as a whole and particularly young people (as I was then) were caught in the quagmire of our mistaken adventure in Vietnam. Patriotism was not popular then. Still, when Armstrong and Aldrin landed on the moon, I remember feeling proud that we had done it, proud that the collective efforts of American society led us to some place that was definitely worth going to.

The Norwegian explorer Fridjof Nansen captured this sense of pride in describing the efforts of his countryman Roald Amundsen, the first person to reach the South Pole:

> . . . The mists were upon us day after day, week after week—the mists that are kind to little men and swallow up all that is great and towers above them.
>
> Suddenly a bright new spring day cuts through the bank of fog. There is a new message. People stop again and look up. *High above them shines a deed, a man.* A wave of joy runs through the souls of men, their eyes are as bright as the flags that wave around them.
>
> Why? On account of the great geographical discoveries, the important scientific results? Oh no, *that* will come later, for the few specialists. This is something that *all* can understand. A victory of human mind and human strength over the dominion and powers of Nature, a deed that lifts us above the grey monotony of daily life. . . .[3]

Pride, however, is a fragile reed on which to hang a sustained exploratory program. Both individuals and societies take pride in getting somewhere first. Nansen, after all, was waxing eloquent about the person who was the first to get to the South Pole. Once the dramatic first journey has been accomplished, interest wanes quickly. After Amundsen reached the Pole, and the Englishman Robert Scott died in a simultaneous attempt, Antarctic exploration was put on ice for years, and human activities on the frozen continent have probably never received the same degree of attention that they did in 1911.

The Apollo program provides an excellent example of the way that interest wanes once a milestone has been passed. A total of six Apollo expeditions landed on the moon, each becoming more sophisticated. The

American public response was a collective yawn. Even now, the Apollo 11 expedition has been relegated to history, far from the general consciousness of many people, particularly young people who didn't experience the events directly. I recently showed some videos of the Apollo 11 expedition to a group of college students. I was surprised to find that this expedition was not familiar to them, and pleased to see them experience the same sense of thrill, of pride, in seeing the old videos of astronauts walking on the moon for the first time.

The three photographs presented here can, perhaps, evoke the same sense of pride and accomplishment as the Apollo videos do. The first, a telescopic photo of the moon, shows an apparently desolate landscape, with nothing but craters and rubble. The telescope shows a bit more detail than the eye can see alone, but the image is still rather familiar. It's almost impossible to imagine people actually landing on the lunar surface.

The other two pictures show that we did, in fact, walk on the moon and drive wheeled vehicles, lunar rovers, over it. A familiar picture from the Apollo 11 expedition shows astronaut Buzz Aldrin on the lunar surface, an American flag patch barely visible on his left shoulder. The next picture shows the lunar rover from the Apollo 16 mission. Astronaut's footprints and tire marks are all over the lunar surface in the vicinity of the lander.

The Apollo expeditions are history. While they can evoke pride in our past accomplishments, it's hard to use this past to kindle the same sense of enthusiasm for a current program. Space planners who want to establish a lunar base face an understandable reaction from skeptics: "You want to go to the moon again? We've been there already." Sure, a lunar base would be bigger and more permanently inhabited, but such details are lost on someone whose sole exploratory motive relies on the exhilaration of getting somewhere first.

Contemporary explorers and adventurers on the earth face a similar difficulty both in finding new things to do and in generating support for their exploits. Now that Everest has been climbed and various improbable milestones like human-powered flight have been passed, travel adventures seem to consist of finding a tiny gap in the *Guinness Book of World Records* and meeting the challenge. Pam Flowers, a woman seeking a challenge, undertook a dogsled voyage from Canada to the North Pole, a route never traveled before by a woman. She had to make this trip alone because she could find neither partners nor financial backing.[4] As it turned out, this trip was the first solo voyage to the North Pole, but it still only

A telescopic photograph of the moon, showing circular dark basins (called "mare" or seas) and circular impact craters. (Lick Observatory, University of California)

made the middle pages of comprehensive newspapers like the *New York Times* and was virtually unnoticed by the rest of the media.

If we can only appeal to pride, it's difficult to kindle an Apollo-like sense of excitement among the public for further space ventures. The only identifiable goal like a lunar landing is a human landing on Mars. But if pride alone is the reason to go to Mars, interest in space exploration will collapse once we get there. The journey to Mars will lead us to the sands

Astronaut Buzz Aldrin on the lunar surface. His suit has an American flag on his left shoulder. (NASA)

Apollo astronauts have landed on the moon, and their feet and the moon rover leave tracks in the lunar soil. (NASA)

of the red planet, but nowhere beyond. Pride can only be used to kindle interest in the early journeys of any venture of exploration. Something else is needed if the program is to continue.

CURIOSITY

One of the chief reasons to explore is to find something new and different. The human ability to share the explorers' thrill, even vicariously, in new experiences is an essential part of the exploring spirit. Throughout the history of human exploration of the earth, expedition leaders have described in loving detail the joy of discovering new plants, animals, and human communities on their expeditions.

More new experiences probably await us in space than in any other human adventure. The experience of being weightless, of floating around in a spacecraft rather than being glued to the seat of a chair, is the most unusual space sensation. It's hard to translate, though, in a way that those of us who have not been in space can understand. The film "The Dream is Alive," projected on a 30- to 40-foot screen, does the best job I've seen of giving an earthbound viewer the sense of free floating. It's possible to buy a number of videos of some footage from the space program. The smaller TV screen provides some limitations, but the sensation of floating is still there.

Gravity doesn't vanish in space. Gravity is a universal force which extends throughout the universe. When you sit in a chair on earth, gravity pulls you toward the center of the earth. Gravity has no effect on the chair since its tendency to pull the chair toward the center of the earth is opposed by the mechanical rigidity of chair, building, and the earth itself. Thus if you were to push yourself up out of the chair, gravity would pull you back down toward the seat, because your body responds to the force of gravity and the chair doesn't.

Now visualize this chair in space, orbiting around the earth. The same force, gravity, which keeps the chair (and the space shuttle or station which it is attached to) in orbit acts in the same way on objects inside the space shuttle. The shuttle, the chair, and you are all falling around the earth in the same orbit. Were you to move upward from the chair, your orbit would change just a wee bit, but the gravitational force on you and on the chair would remain the same. There would be nothing to pull you down toward the seat.

Some of the experiences of weightlessness relate to the difference in carrying out ordinary tasks. Were you to be in space and let go of this book, it would remain suspended in air. If you gave it a tiny push in letting go of it, it would float upward or to one side. If you wanted to read some more, you would have to fetch it from wherever it had wandered off to. Finally, the air currents set up by the ventilating fans in your space environment would probably have turned a few pages in the book for you, and you would have lost your place.

It all sounds rather fanciful, but it really happens, and it makes some difference in the way that astronauts live and work. Think about something as simple as using a book, for example. Like pilots and many other people, astronauts rely on checklists written in notebooks, referring to the books to see what has to be done next. Instinctively, an astronaut might put a notebook down on a table, only to find that it floated away, maddeningly out of reach. Once she finished adjusting or checking whatever it was that she was supposed to adjust or check, she would then have to go and get the book, find her place in the checklist again, and work on.

Decades of experience in the space program have eased some of the practical difficulties of functioning in weightlessness. The interior of space vehicles is covered with little bits of Velcro, which can keep similarly equipped tools, books, and checklists from floating away. Toilets have become better designed, but are still a source of difficulty. What's still most unusual is the sensation of being in a weightless environment. I've never been in zero gravity, and so I'll let Apollo 11 astronaut Michael Collins describe the experience:

> It is a strange sensation to float in the total darkness, suspended by a cobweb's light touch, with no pressure points anywhere on my body. Instinctively, I feel I am lying on my back, not my stomach, but I am doing neither—all normal yardsticks have disappeared, and I am no more lying than I am standing or falling.[5]

Looking at the earth is another memorable aspect of the space experience. As the illustration shows, you can see much more of the earth than you can from an airplane. Entire storm systems, like the Indian Ocean hurricane in the picture, become visible outside of one window. Astronauts report that in what little free, unscheduled time they have, earthwatching is the most popular pastime.

The space experience is, then, profoundly different from being on the ground. I suspect that if a large fraction of the American public could actually enjoy this experience firsthand, support for the space program would be much more substantial than it is. However, only a hundred people have ever been in space, and the rest of us, including myself, have to rely on reports from astronauts, videos, and pictures. Unfortunately,

Hurricane Kamysi in the Indian Ocean is shown in this photograph from Space Shuttle mission 41-C, conducted in 1984. Earth-watching is one of the most exhilirating of space experiences. (NASA)

NASA has not managed to share the space experience with the public in the same way previous explorers have. The videos and pictures from these space missions are spectacular, but they are not broadcast too widely. They gather dust on NASA shelves; you can buy them from private companies if you know where to write.[6]

A difficulty that many of us have in appreciating the space experience is that astronauts have generally been selected for their skills as test pilots or scientists, not for their ability to share their experiences with others. Many astronauts, indeed, were quite uncomfortable with the rounds of speeches which they had to endure; Buzz Aldrin is quite frank in his book about the physical discomfort which he often endured during the course of preparing and giving a speech.[7]

Perhaps as a result, the literature on the space program written by the astronauts themselves is quite limited. In the course of writing my books on the space program, I have spent some fair amount of time in the University of Delaware library seeking books on various heroic human exploratory expeditions. The literature written by participants is vast. Naturalists and painters went on many expeditions for the express purpose of creating something which could be shared with others. In contrast, the literature on the space program is mostly written by nonparticipants. There are precious few books which manage to recreate the astronaut's experiences. Joseph Allen's *Entering Space: An Astronaut's Odyssey* is one superb oasis in the desert of books by participants in the space program. Congressman Bill Nelson, one of the first "civilians" to fly on the Shuttle, has just published a book about his flight on the last mission before the *Challenger* disaster. For younger readers, Sally Ride and William Pogue have provided a more practical view of what it's like to live in space.[8]

NASA began an effort to reach out to more people with the Teacher in Space program, in which Christa McAuliffe was to share her experience with schoolchildren, teachers, and anyone else who would listen. While McAuliffe's tragic death in the *Challenger* explosion was a severe setback to the Teacher in Space program, it does continue. McAuliffe's backup, third grade teacher Barbara Morgan, is ready to fly anytime. In each state, the two finalists in the Teacher in Space program continue to visit classrooms, using space as a way of sparking student interests.

HEROISM

Pride and curiosity are two aspects of the exploring spirit; a third is heroism. The tragic death of the *Challenger* astronauts reminded all of us that space exploration is a dangerous human adventure. Prior to *Challenger*, NASA's press releases describing the space shuttle as a "space truck," and posters that proclaimed that the agency was "going to work in space" were an attempt to demystify space travel that was both ill-conceived and incorrect. Being hurled into orbit atop a pillar of flame and floating weightless with the earth out the window below is and always will be rather different from driving a Mack truck down an interstate highway. It is much more exciting—that's why a job as an astronaut is harder to get than a job as a truck driver. And, as *Challenger* reminded us, space travel is also dangerous.

Heroes are brave people. But how brave? What is bravery? We know little of how the *Challenger* astronauts reacted to their tragedy. Commander Dick Scobee's last transmission to the ground, "Go at throttle up," was made well before any obvious signs of trouble. Shuttle pilot Michael Smith's last words were recovered from a tape of the words spoken over the shuttle intercom. His enigmatic "Uh-oh," spoken only a split second before the explosion, could be a heroic acceptance of impending disaster, or it could simply reflect his ignorance of what was going on.

We do know how the *Challenger* astronauts' spouses and associates reacted to the tragedy, and what they did. Heroes accept the risk of death—or the risk of bereavement for their spouses—as part of the cost of going into space. One spouse's continued support for the space program is testimony to her bravery, her acceptance of the *Challenger* tragedy as part of her husband's job. June Scobee, widow of mission commander Dick Scobee, has devoted her time, energy, and fame to found the *Challenger* center, an effort to publicize the space program and to establish closer contact between the space program and elementary school teachers. This is not the act of someone who seeks satisfaction in blaming NASA or the rocket manufacturers for the accident which tragically upset her life.

The response of the other NASA astronauts to the *Challenger* tragedy was typically heroic. Rather than avoid the space program, or move to a less demanding job, astronauts continue to jockey for opportunities to get into space. Those astronauts who have left the space program seem to have left because their opportunities to ride the space shuttle will be more limited in the 1990s, not because they don't want to go up. There were some initial complaints from the astronaut office that NASA was not giving them enough say in various mission decisions, but these complaints are not signs of cowardice. Heroism doesn't mean blind acceptance of a ludicrous policy set from above. Rather, it is an acceptance of risk, a willingness to do whatever is necessary to meet a challenge.

Antarctic Heroism: Staring Death in the Face

It's often hard to fathom the heroic attitude toward death. Before (and sometimes even after) a dangerous venture, heroic explorers are generally reluctant to talk about their feelings about the dangers; a stoic acceptance of risks usually carries with it a reluctance to share one's emotions. Author Tom Wolfe had some success in interviewing astronauts and their associates for his memorable book *The Right Stuff*[9]; he produced a vivid portrayal of understated courage as an integral part of an astronaut's personality. However, even a book like Wolfe's can only investigate how heroic people think they'll react when faced directly with impending death, since a direct record of their thoughts is often hard to come by. Explorers tend to die quickly, often in isolated places with few other friendly humans around, and in ways that are quite inconvenient to authors and historians who want to know just what the right stuff really is. Even if they had thought to record their last thoughts on paper or on a tape recorder, the notebooks or audiotapes vanish into the briny deep when a ship sinks in a storm, or are shredded into a billion pieces when an experimental aircraft crashes in the California desert.

The English Navy captain Robert Scott provided us with one of the few insights into the thoughts of perishing heroes. Scott died in his attempt to reach the South Pole, but he continued to write in his diary even as he and his companions lay trapped in a tent, less than a dozen miles from their next supply base and salvation. Scott's journal, found in his tent after he died, contains some guide to the true ingredients of the right stuff.

The heroic age of Antarctic exploration culminated in the race to get to the South Pole in 1910. The Norwegian Roald Amundsen and the Englishman Robert F. Scott both set off for the pole in October of that year. Amundsen, who had dedicated his life to Arctic and Antarctic exploration since his teens, prepared the way well, laying out a whole string of supply depots during the southern hemisphere summer of 1910–1911. They waited and built up their strength through the long Antarctic winter. Amundsen, four members of his expedition, and his dogsleds reached the pole on December 14 and returned to their base on the edge of Antarctica on January 25, 1912.

The less fortunate and worse-prepared Scott didn't make it. The words in his journals transform a difficult journey into what Stephen Pyne calls a "great moral drama"[10] that captures the essence of heroism. Even after he saw that Amundsen had reached the pole first, he persisted in his journey, reached the pole, photographed Amundsen's tent, and then turned around to race for home before winter closed in. By mid-March 1912, Titus Oates, one of the five people to trek to the pole, had weakened with frostbite so badly that he couldn't walk on, much less pull a sled. He was slowing his companions down, potentially endangering the entire expedition.

What was to be done? Oates asked Scott to leave him in the tent to die. Scott refused, and the party limped on. Oates himself made the final decision. On March 16 or 17 (Scott lost track of dates) Scott's diary describes what happened to Oates:

> We can testify to his bravery. . . . He [Oates] woke in the morning—yesterday. It was blowing a blizzard. He said, "I am just going outside and may be some time." He went out into the blizzard and we have not seen him since. . . . We knew that poor Oates was walking to his death, but though we tried to dissuade him, we knew it was the act of a brave man and an English gentleman. We all hope to meet the end with a similar spirit, and assuredly the end is not far.[11]

A few days later, Scott and his two remaining comrades found themselves trapped, only 11 miles from the last supply depot which lay between them and Scott's base on the Antarctic coast. In good weather, 11 miles was only one day's travel. But move they could not. A nine-day blizzard trapped them in the tent, where they grew progressively weaker. Eventually, they realized that they would not be able to make it. In his final days, Scott used his waning energies to compose a dozen letters to backers of the expedition and to relatives of the explorers who died with him. Some excerpts follow:

> If this letter reaches you, Bill and I will have gone out together. We are very near it now and I should like you to know how splendid he was at the end—everlastingly cheerful and ready to sacrifice himself for others, never a word of blame to me for leading him into this mess. [to Mrs. E. A. Wilson]
>
> It seems a pity, but I do not think I can write more.
>
> Last entry. For God's sake look after our people.[12]

Is this Tom Wolfe's "right stuff," what heroes are made of? You bet it is.

Apollo 13: Almost Lost in Space

The third scheduled Apollo landing flight provided another example of calm acceptance of danger. When this mission was on its way to the moon, one of two oxygen tanks exploded, dumping its contents into space. Two of the three fuel cells providing power to the spacecraft stopped working. Within a matter of hours, the command module, where the astronauts were supposed to live and work on the way to the moon and back, had to be completely shut down. Fortunately, the astronauts had a "lifeboat" along—the lunar module, intended to take two of them down to the surface of the moon.

A successful rescue of the Apollo 13 flight required a reasonably calm reaction to the disaster, both on the part of the astronauts and on the part of the flight controllers in Houston. A spacecraft like Apollo is quite complex, and repairing it requires a detailed, intricate analysis of all of the interacting components. Mission plans, usually developed over a matter of months, had to be reworked in a few hours.

Journalist Henry Cooper later wrote that the astronauts were almost too calm. "Houston, we've got a problem," Command Module pilot Swigert reported, so quietly that Jack Lousma, in Houston communicating with the astronauts, could not recognize his voice. The explosion set the spacecraft tumbling, and Captain Jim Lovell patiently adjusted the spacecraft's tiny rocket thrusters to bring it under control. Factual reports were relayed up to the astronauts and back down to the ground; the voices of flight controllers remained level as they talked to each other over various loops of telephone linkups which kept them in contact with the right people.[13]

Later on, other problems developed. The machinery in the Apollo spacecraft is sufficiently interlinked that a power shortage also limits the amount of water that the astronauts could use. The astronauts had been severely rationing their own water intake, allowing themselves a pint and a half of water for four days, five times less than the normal daily minimum requirement. They said nothing about this idea to the ground, stoically accepting the necessary inconveniences (but making it harder for the doctors on the ground to diagnose their problems).

While the astronauts remained worried, checklists were read up to them and read back down in calm voices, as though they were no more important than a list of items to pick up at the store on the way home from work. "O.K. Panel 250, circuit breaker battery A power entry, and post-landing closed."[14]

* * *

Thus there is an exploring spirit in human beings, a combination of pride, curiosity, and heroism, a sense of adventure which has been the driving force behind a number of expeditions. Is it enough, by itself, to motivate a long, sustained program of exploration of the globe or of the solar system? If astronauts alone could make a space program, the answer would undoubtedly be yes.

However, the two examples of exploration discussed in this chapter suggest otherwise. While pride and power fueled the American effort to get to the moon, public interest waned quickly once we had reached the finish line and left footprints in the dust at Tranquility Base, the landing site of Apollo 11. The grand moral drama of the race to the South Pole was followed by a 30-year hiatus in Antarctic exploration. While the exploring spirit is a powerful human motive, it has its limitations.

The question for the future is what kind of exploratory ventures can generate enough political, social, and economic support in order to succeed. If the exploring spirit by itself is not enough to support something like a sustained space program, what else is needed? The historical analog to the space program is generally seen as the Great Age of Discovery in which many household names like Columbus and Magellan were the grand heroes. In the next chapter, I will consider the space program in the context of the Columbus expedition and see what we can learn about the nature of our political will to support exploratory ventures.

CHAPTER 3

Exploration

A Human Imperative?

From the very beginning of the space program, its proponents have sought to evoke the Golden Age of Discovery, that momentous era when humanity, led by Western Europeans, became aware of the true dimensions of the globe. Many satellites and missions have names that evoke this era. The first American satellite was called Explorer 1, and the name was given to a whole series of small scientific satellites. The two most famous explorers, Columbus and Magellan, have posthumously given their names to a future mission to Venus and to the European part of the American space station.

Does our present effort in exploring the solar system really represent the beginning of a modern Golden Age of Discovery? Does the dramatic impact of the Golden Age of Discovery on human history indicate that exploration and settlement of the solar system is our human destiny? Can the exploring spirit, by itself, provide the political will necessary to sustain an ambitious space program?

The most enthusiastic proponents of space travel—people who foresee colonization and settlement of the inner solar system—answer all of these questions with a resounding yes. In this view, the space program will be seen as historically inevitable, the manifest destiny of a nation which was born in a cradle made by Columbus, Magellan, and others. If astronauts Yuri Gagarin, Neil Armstrong, and Buzz Aldrin are no more than the contemporary equivalents of these legendary explorers, then those who don't enthusiastically support the space program should easily be persuaded to change their minds with a simple appeal to the heroes we all learned about in elementary school.

Were the space program our manifest destiny, NASA's budget problems would evaporate instantly. In my view, which is shared by many space program supporters as well as skeptics, our reasons for going into space are more complex. Space scientist James Van Allen overstates the case, but is basically correct in claiming:

> Fervent advocates of the view that it is mankind's manifest destiny to populate space inflict a plethora of false analogies on anyone who contests this belief. At the mere mention of the name of Christopher Columbus they expect the opposition to wither and slink away. If reference to Columbus is made in an offhand, thoughtless way, it is incompetent; but if it is made with full knowledge of the facts, it is deceitful and fradulent.[1]

What are these facts that Van Allen talks about? What sort of space program do we and do other nations have the political will to support? What does it take to develop the popular and legislative support which is required if any nation—be it the United States, the Soviet Union, or someone else—is to sustain the momentum which can propel humanity into outer space? This chapter will touch on various episodes in the history of exploration in an attempt to answer this question. I make no claim to write a complete history of human efforts to expand our horizons; rather, this is a random walk through the history of exploration in which I tell various stories in order to demonstrate that some sort of practical payoff has been necessary in order to justify large-scale exploratory ventures like the space program.

EXPLORATION: THE NECESSARY PAYOFF

Americans, Europeans, and leaders of the Soviet Union will be making various decisions in the next decade to determine the future of their space programs. The political and social realities which govern the collective behavior of these groups of citizens will determine just what kind of space program is supportable in the long term. In centuries past, organized social groups, some of which are not even nation-states in contemporary terms, have devoted some of their resources to exploration in the past. Why did they go? What were the obstacles that had to be overcome? Who supported the cost of their expeditions? What can we learn about the exploring spirit from a historical retrospective?

The picture will emerge in the pages to come, but let me provide some broad outlines here. Yes, indeed, there is an exploring spirit which stirs human hearts. However, Columbus and Yuri Gagarin, the first man to orbit the earth, could not have completed their historic journeys without backing from someone else, and it is very rare that the exploring spirit alone can generate the political and financial support needed to make these exploratory ventures actually happen. The First Great Age of Discovery, the famous first two centuries of exploration, was a buccaneering era in which financial profit (supplemented by religious imperialism) was the driving force which led to the exploration of the globe by Western European mariners. While the exploring spirit moved Columbus, his backers were basically interested in getting rich.

Indeed, the promise of commercial return sustained much of the interest in the space program in the 1950s, when it was first getting started. It's widely agreed that one of the most important milestones in the development of space was the publication of a rather far-sighted article in the obscure radio magazine *Wireless World* by a then obscure Royal Air Force communications officer, Arthur C. Clarke. Clarke pointed out that orbiting satellites could be the keystone of a global communications system, since one satellite which orbited the earth every 24 hours could receive and send signals from a large part of the earth's surface, serving as a very powerful relay station.[2]

Another driving force behind the space program in its early days has reemerged through former President Reagan's "Star Wars" program: the military. In the 1950s, the military's need was for satellites which could take pictures of the Soviet Union and open up the closed society. In the 1980s, some people within the military captivated the attention of an aging President and a technologically naive public with visions of an "astrodome defense" which could ward off an enemy's nuclear-tipped missiles in the same way that the roof of the Houston Astrodome keeps rain off of the Houston Astros baseball team. Scientists inside and outside the Strategic Defense Initiative program now agree that the idea of a system which could defend against 10,000 nuclear warheads coming from a nuclear superpower like the Soviet Union is a technological fantasy. However, some kind of defensive system which could cope with far more modest attacks might be technologically feasible and might even make sense in the 21st century, if the nuclear world of the 21st century is different from the one which we now know. It's quite possible that a century from now, the superpowers will have far fewer warheads than they do now, and that a system which could defend against accidental missile launches and against threats from nuclear minipowers might make sense.*

However, neither communications satellites nor spy satellites nor the Star Wars program can justify settlement of the inner solar system, or even

*There's more discussion of Star Wars in chapter 13.

establishing space stations in orbit near the earth. None of these ventures requires the human presence in orbit, for they all rely on automatic machinery. Space enthusiasts who advocate settlement of the solar system must rely on something else.

In the waning years of his administration, President Reagan announced a formal space policy which included the expansion of human presence beyond earth orbit as a major policy goal.[3] However, a presidential announcement does not automatically mean that the nation will line up behind such a policy and that the necessary political consensus will appear. Such a consensus did not, in fact, emerge by the time the Reagan administration left office, and it remains to be seen whether the necessary support for a long-range program of space settlement will arise.

What will it take to develop such support for an expansionist space program? One way to answer this question is to take a look at history, recalling Santayana's famous aphorism that those who are ignorant of history are compelled to repeat history's mistakes. What did in fact propel the Great Age of Discovery? In particular, what was needed in order to generate political and financial support for these ventures? Why did the people who provided the money needed for these expeditions invest in them? Unless human nature changes, we must expect that the reasons then will be the same as the reasons now.

The historical picture is, as history often is, rather complex; it's not just a matter of dates or simple labels. What does emerge from a treatment of the historical record is that the exploring spirit alone was not enough to generate support for Columbus, Magellan, and their contemporaries and successors. Capitalistic payoffs, return on investments, a favorable bottom line, to cite a number of labels, were absolutely crucial in convincing the political systems of the time to support Columbus. Even subsequent expeditions, which paid more attention to scientific exploration rather than buccaneering exploitation, had practical motives such as the mapping of strategic territory or establishing a human presence in places with potential strategic importance.

THE FIRST, BUCANEERING AGE OF DISCOVERY

Christopher Columbus is widely cited as the precursor of the current generation of astronauts. Columbus was an expert navigator and a genius at persuading his crew to sail off into the open ocean and to stay on course despite the potential difficulties of getting back. His reason for going, though, was not just that his soul contained the essence of the exploring spirit. He also had a chance to make money by finding gold or spices, bartering for them at low prices, and reselling them on European markets at an enormous markup. These kinds of financial opportunities only exist today in a few lines of business: selling illegal drugs like cocaine or volatile but valuable pieces of paper like the stocks of companies that are takeover candidates. The financial side of the Columbus expedition is not terribly heroic, but too bad; it mattered then and matters now.

Certainly, Columbus is a towering figure, a colossus of the First Great Age of Discovery. Anyone who can set sail into a supposedly empty ocean has tremendous courage, as well as a good healthy dose of the exploring spirit. His persistence was indomitable—his vision of the "Enterprise of the Indies" remained alive for ten years while Spanish, Portuguese, and other royalty kept rejecting his pleas for sponsorship. He was a superb seaman, without peer among the many great explorers who set the stage for Magellan's historic voyage around the world. His desire to get rich from his discoveries doesn't detract from his well-deserved place in history.

The legend that Americans learn in grade school is that Columbus came up with this idea of sailing west to reach Asia, persuaded Queen Isabella of Spain to support him, set forth, and discovered America. To his death, he thought that the islands he discovered were part of the Japanese archipelago, destined to become the exurbia of Tokyo at some future date. For many, Columbus's motives are obscure and irrelevant. It is legends like these that perpetuate the misconception that there is some exploring spirit which was enough to motivate Columbus, with no need for thoughts of money or fine titles.

A slightly deeper version of the legend says that Columbus sought "spices" in his quest for a route to Asia. However, this motivation has always seemed rather suspicious to many Americans; at least it did to me. Spices are rather cheap these days. Going halfway around the world to fill an extra can or two on the supermarket spice rack seems, perhaps, unbelievable. It is tempting to view the quest for "spices" as some kind of historical camouflage, and attribute some romantic, existential exploratory urge to Columbus. Many writers do.

Such a perception is incorrect, and the historical record clearly shows it. Columbus's deal with Ferdinand and Isabella did give him the title of "Admiral of the Ocean Sea" and a 10% cut on all trade with the East, among other things. (Incidentally, Columbus's heirs never received their 10% share; his legitimate grandson and daughter-in-law reached an out-ofcourt settlement with the Spanish crown some decades after Columbus died.)

These commercial motives are an even more important part of the decision by Ferdinand, Isabella, and several other Spanish grandees to back his venture. They sought profit from the spice trade, which was a fabulous source of wealth in 15th-century Europe. Their investment in his journey was relatively small. Why in the world were spices so important and so valuable at that time?

Spices: The High-Priced Drugs of the Renaissance

Spices in the Renaissance were like illegal drugs are today: *very* valuable cargoes. One camel load of spice was worth a half million 1988 dollars; a shipload was worth even more. One individual who could obtain one of these cargoes and resell it at a modest profit (say 30%) was able to make a good year's living from one single transaction. Because spices were, at the time, only grown in India and points east, a few people who happened to be living in the right places were able to monopolize and take considerable advantage of this trade. The herbs which can be grown in Western gardens were (and are) less pungent than the peppery stuff which requires a tropical climate to grow in; in addition, they weren't as well known then as they are now.

But why would Europeans want to buy these expensive spices in the late Renaissance? You have to go back a bit in time to recreate the great attraction of these high value substances. For nearly 300 years, from the 8th to the 11th century, Western Europeans simply could not buy spices at any price, because the explosive expansion of Islam made much of the Mediterranean into a Moslem-controlled lake.[4] They ate bland mush, stews made from heavily salted meat and tasteless root vegetables, day after day. There was no tea, coffee, or tobacco, no lemons or oranges. I doubt that there are many Americans who do not enjoy at least one of these flavors every day, particularly considering that caffeine is one of the essential ingredients in Coke and Pepsi.

In the Crusades, hordes of Christians tramped from Northern Europe to what is now Israel in an effort to wrest the holy places of Jerusalem from their Moslem occupiers. The Crusaders ate tasty food for the first time, and realized that small quantities of good stuff could transform bland stews into gourmet meals. While the Crusaders eventually were beaten militarily, they brought back spices to Europe, creating a market for expensive tastes. They had a profound effect on the cultural, culinary, and economic life of Europe.

Spices, particularly nutmeg, pepper, and cloves were also thought to have some kind of medicinal or preservative value. The doctors of the day prescribed bad-tasting concoctions of herbs and spices.[5] In Northern Europe, farm animals could not easily survive the winter, and the standard practice was to slaughter many cows and sheep in the fall, salt the meat, and eat it through the winter. (There were no refrigerators, of course.) At the very least spices could disguise the taste of spoiled meat, and they might well slow down bacterial action a little bit.

So there was no problem in marketing spices. How much were they worth? Take an example. Sebastian Del Cano, leading the wretched remnants of Magellan's expedition back to Spain in September 1522, brought back 520 quintals (about 55,000 pounds) of cloves, which fetched a price of 7,888,634 *maravedis*. In purchasing power terms, it turns out that a *maravedi* was worth somewhere between 50 cents and $1, according to sources which range from Cervantes's classic *Don Quixote* to old journals. I'll adopt a mean figure of 75 cents = 1 maravedi and round off the modern equivalents to avoid creating the illusion of precision.[6] (If you want to know the gory details of how I arrived at this figure, go to the back of the book and read the indicated footnote.) This meant that one ship brought back about $5 million worth of cargo, more than enough to pay for the Magellan expedition.

The backers of the Magellan expedition realized 143 *maravedis* (or about $100) for each pound of cloves that the expedition brought back, roughly a hundred times the current wholesale price.[7] Is such a price reasonable? A German price table of 1393, in which the value of one or two pounds of various types of spices is set equal to the value of a draft animal like a horse, a cow, or an ox, confirms my spice price estimate of a $100 per pound.[8]

Big Bucks from the Spice Trade—Back Then

A small ship, carrying about 50 tons of spice, carried about $10 million worth of cargo. The only trade like it today is the illegal cocaine business, where one car trip between Florida and New York can keep a person alive for a year (unless he is addicted to the stuff and has a rather high cost of living). No wonder piracy was so common and so profitable.

In principle, competition between various middlemen or various trade routes might be able to restrain markups in the spice business, but geography and history conspired to create a Moslem monopoly in the century before Columbus. The spices, most typically pepper, cloves, nutmeg, and mace, came from plants which could only grow in wet, tropical areas, which at the time meant India and the islands east of it. Political chaos in central Asia and the fall of Constantinople in 1453 left only one route from the spice islands to European markets: by sea in the Indian ocean to Egypt, then across Suez, then in Turkish or Venetian ships to European market cities.[9] The Spanish and Portuguese, at the tail end of a long supply line, had ample commercial reasons to seek a shorter trade route which they could control (and earn profits from). Signs of profits from newly discovered lands appeared even before Columbus, when Portuguese who mapped the east coast of Africa traded bits of cloth for small quantities of *malagueta*, a peppery spice which could be used as a substitute for black pepper, and bits of gold, which Africans obtained from a mysterious source in the Dark Continent's interior.[10]

To be fair to Columbus and his successors, I should at least mention their secondary, purer motive for exploration. Bernal Diaz, one of Cortez's footsoldiers who participated in the conquest of Mexico in 1519, put it clearly. He wrote that he and his friends went to Mexico "to serve God and His Majesty, to give light to those who were in darkness, and to grow rich, as all men desire to do."[11] The Spanish "gave light to those in darkness" by building a chain of missions in Mexico, Arizona, and California and converting the Indians to Christianity. It's unlikely that there are convertible beings in space, and so the missionizing aspect of the Great Age of Discovery is no precedent for the space program.

A latter-day Columbus seeking to settle the inner solar system and make some commercial return from it would then have to find something like gold or spices on the surface of the moon, on the deserts of Mars, or in the asteroids. Perhaps something like that exists, perhaps not. But if Columbus is to be seen as the predecessor of today's space explorers, various countries' space agencies must focus considerably more attention on the commercial return from the space program than they have done so far.

THE SECOND GREAT AGE OF DISCOVERY

If Columbus, Magellan, and their successors were not the spiritual ancestors of today's astronauts, who was? The buccaneering age of exploration, when a single expedition could by itself create enormous wealth, only lasted for a few centuries. From, say, 1400 to 1700, instant riches, in the form of gold or exclusive trade routes, were easy to come by. Later on, backers of an exploring expedition might be amused by tales of the South Pacific, with its friendly natives, but such ventures did not produce the same kind of tangible return on their investment.

Historian William Goetzmann has identified a change in the motives for exploration during the 18th century. Explorers sought knowledge for its own sake, not gold or spices. Frequently they were on governmental payrolls, producing reports to governments and to people rather than shiploads of valuable cargo.[12] Goetzmann has argued that this is a "Second Great Age of Exploration." The heroes are not the legendary figures whom every American elementary school student has heard of; rather, they are people like Alexander von Humboldt, James Cook, and Matthew Maury.

These names are not exactly household words. Their historical obscurity illustrates the difficulty of maintaining sustained public interest in explorers who are merely discovering new rivers or mountains rather than new continents. A close look at their ventures reveals not a continuous pattern of exploration, unified by a common motive, as was the First Great Age of Discovery, but a patchwork of different expeditions, generally supported by individual departments in different governments, each with their own motives, many of them quite practical.

While historians identify the Second Great Age of Discovery with such human pursuits as "science" and "knowledge," there were a number of pragmatic motives as well. In America, government support of science has always had to justify itself in the name of something practical like railroad building, agriculture, or industrialization. Today, science and education are being sold as the keys to economic competitiveness. I'm a space scientist, and I would love to portray the Second Great Age of Discovery as the precursor of a "research only" space scenario, in which knowledge for its own sake was the sole reason for voyaging. Such a portrayal would betray the truth of history.

Goetzmann identifies this "Second Great Age of Discovery" as beginning with the voyages of Captain Cook in the 18th century. Cook's first expedition had several motives, most prominently the potential discovery of a large southern continental landmass. For millennia, geographers had, on esthetic grounds, believed that this continent must exist in order to balance the elephantine mass of Europe and Asia in the north. Such a continent could contain the usual complement of exploitable natives, gold, and fish, even if it was too far to the south to grow spices. One of Cook's objectives on his voyages, secretly transmitted to him by the British Navy, was to discover this southern continent and claim it for the British Empire. So much for pure research.

A generation after Cook, the Prussian explorer Alexander von Humboldt captured, more than anyone else, the spirit of the Second Great Age of Discovery. Humboldt was born rich and paid for his own expeditions, spending his entire fortune on them and dying impoverished. He didn't need to generate popular support, but a Humboldtean approach to space exploration is impossible. The richest American could keep NASA going for only a month. Humboldt was fortunate enough to live in an era where one individual could master, to some degree, every field of science and contribute to them. His multivolume work, *Cosmos*, ranged from the origin of the solar system to the interrelationship of the earth's oceans, atmospheres, landscapes, and inhabitants.

While practical applications might seem far from Humboldt's botanical, geographical, and artistic interests, there were many ways in which the knowledge gained in the Second Great Age of Discovery had immediate practical applications. Matthew F. Maury, a naval officer, founded the science of oceanography because an accident made him unfit for sea duty. Reluctantly stuck at a desk in Washington, Maury pored over naval logbooks, organizing thousands of observations of wind and current direction made by sea captains over the years. He was the first to realize that you could use these data to ask, for example, how likely it was that a boat south of Cape Horn would find the wind blowing from the east at some time of the year. (South of Cape Horn, between South America and Antarctica, the wind usually blows from the west.) When his charts were published, sailing time from San Francisco to New York was slashed by a third, from five months to three, with considerable commercial benefits.

THE ANTARCTIC

The culmination of the Second Great Age of Discovery was the discovery and exploration of Antarctica, a diminutive, ice-bound, uninhabitable landmass at the bottom of the world. The opening of the Antarctic is often cited by space enthusiasts as an example of a sustained program of scientific exploration, with exploitation supposedly playing a very small role. But as we shall see the exploration of the Antarctic is not just pure science. There is at least some commercial potential in the Antarctic; indeed, various nations have been obtaining large quantities of fish and seals from the Antarctic for centuries. The same curious synergy between science and commercial exploitation that enabled Matthew Maury to found the science of oceanography and simultaneously give the American merchant marine a competitive edge drives scientists, explorers, and governments to establish bases on the southernmost continent. As a result, Antarctica has been the site of a fascinating experiment in international politics: a continent administered cooperatively by several nations, not taken over by one colonial power.

The exploration and establishment of bases in Antarctica set some important historical precedents for space exploration. The curious blend of economic, nationalistic, and scientific interest in the frozen continent has persisted through most of the space program. Governments have sent expeditions to the Antarctic without the prospects of immediate economic return, and also with no possibility of establishing territorial claims in the traditional sense. However, it would be an overstatement to say that increasing knowledge could by itself motivate Antarctic exploration and justify the expense of sending and maintaining expeditions there.

The Scientists' Continent

Roald Amundsen and Robert Scott raced for the South Pole, and Amundsen won. What was left to be done in the frozen continent? The English explorer Ernest Shackleton attempted to cross the entire continent in 1915; this seemed to be the next logical step in Antarctic exploration. Shackleton scarcely got started; half of his expedition was trapped in ice far from the continent, drifted on an ice floe when their ship foundered, and ended up on Elephant Island, at the tip of the Antarctic peninsula. This expedition did not reach its goal, though Shackleton's journey from Elephant Island to South Georgia Island, in an open boat across 500 miles of some of the coldest, roughest water in the world, was one of the most daring and dramatic ocean voyages which has ever occurred.[13]

With the pole reached, and a continental crossing apparently unattainable, what now? For the next 40 years, a few nations sent sporadic expeditions to the frozen continent. Many of these were privately financed; governments often saw no reason to fritter away resources in a deep freeze. Some expeditions sought to establish territorial claims, anticipating the commercialization of the Antarctic. Norway visualized an enormous whaling industry in the south, and claimed an isolated island and a large, 60-degree slice of the continent. Britain, Argentina, and Chile all claimed the economically intriguing part of the continent, the serpentine peninsula which snakes its way northward toward the tip of South America, the only part of the continent which is not permanently swaddled in impenetrable ice. These conflicting claims led to international conflicts, with shots being fired on one occasion.[14] (No one was killed or even hurt.) Everyone sought his share of the pie, claiming huge wedges of ice between various longitude ranges.

The American pilot Richard Byrd introduced airplanes to the Antarctic in 1928. The expeditions Byrd led were able to map huge portions of the Antarctic with much less effort than people trekking over the ground. Another result of the Byrd expedition was that the old, imperialistic methods of laying claim to a territory were meaningless in practical terms. Airplanes could, and did, fly over the ice and drop national emblems like flags and coins in their path. In 1939 Germany's expedition to "Neue Schwabenland" dropped swastikas to the ice sheet and mounted shore parties to raise the Third Reich's flags on Antarctica. But establishing a practical claim to a territory requires settlement, and settlement of the Antarctic in the same sense that applies to more temperate areas is impossible. Even a dozen isolated scientific stations cannot defend or even delineate a "frontier."

Some nations sought other means to establish territorial claims, particularly after World War II. Stephen Pyne, providing a humanist's view of the continent in his book *The Ice*, mentions several comic ones.[15] An Argentinian woman in the late stages of pregnancy was flown to a scientific base so that the baby could be born on "Argentinian" territory. A Nazi supposedly saluted a penguin with chants of "Heil Hitler"; one wonders what the penguin made of it, and who was more intelligent, the Nazi or the penguin. It seemed that the Antarctic was to go the way of Africa, to be carved up by a number of powers.

The solution to the Antarctic territoriality problem came from scientists rather than diplomats. Scientists dreamed up and implemented a coordinated program of earth and space exploration in 1957–1958, the International Geophysical Year (IGY), the same international program which spawned the space age. In April 1950, a group of earth scientists gathered at James Van Allen's home in Washington to meet Sydney Chapman, a famous, bright, and very energetic English geophysicist. These scientists realized that military necessity in World War II had stimulated the development of vehicles and electronic instruments which would make Antarctic exploration considerably easier. They wanted to use this technology in a large-scale, internationally supported program of polar exploration, similar to two International Polar Years which had occurred in 1882 and 1932. A key component of the International Polar Years, as well as of the IGY, was international cooperation and open exchange of the information which was obtained.

Timing is often critical to the success of any proposal, and the timing of the IGY was perfect. In the Antarctic, the IGY was a framework in which the United States and the Soviet Union could establish a substantial presence on the continent without becoming embroiled in the question of whether or not to stake any territorial claims. Argentina, Britain, Chile, and the all other nations participating in the IGY reached a "gentleman's agreement" to put their legal and political battles on hold for the period.[16] The IGY exercise established the precedent of freedom of travel on the continent and the presence of foreign scientists at any country's scientific bases.

More immediately relevant to the space program was the IGY's role in providing an excuse, a scientific cover, for an earth satellite launch. The same synergy of commercial, military, and purely scientific interests which had pushed Antarctic exploration could come together under IGY auspices and push for the earth satellite project. The military wanted to launch reconnaissance satellites to photograph Soviet military bases from orbit, but a purely military satellite program would have raised some complex international issues of territoriality—does a nation "own" that part of outer space which extends beyond its borders? The telephone company, in the form of scientist John Pierce, knew about the potential of communications satellites, but this potential could not justify the huge investment needed to build a rocket that could propel a satellite into orbit.

As a result, the first American satellite program, the ill-fated Project Vanguard, began as a basically scientific program, part of the IGY, to launch a satellite into space for scientific purposes. No nation could object to a scientific satellite passing over its territory; after all, a satellite is a good idea, and it has to pass over many national boundaries as it circles the earth several hundred miles up. It is simply not possible to put a satellite into an orbit near the earth which only carries it over open oceans, or which keeps it over the territory of one particular nation.

Territorial Claims in Deep Freeze

The aftermath of the IGY had, perhaps, the most significance for space exploration. During the IGY, many countries mounted scientific expeditions as a tangible way of showing their interest in the frozen continent. The Antarctic component of the IGY was extended for another year, and the participants stayed on. It soon became clear that the Russians had no intention of leaving, and if the Russians would stay, so would the Americans. With everyone staying in the Antarctic, some kind of agreement covering the continent's future and dealing with the question of territoriality was needed.

What finally came out of a very carefully conducted set of diplomatic discussions was the Antarctic Treaty of 1959, which essentially froze the territorial claims in the limbo which the IGY had put them in. During the life of the treaty, no new claims can be made, and no claimants are to use their activities in the continent as a way of supporting their own claims or denying anyone else's. Finn Sollie, a specialist in international law, describes the treaty as a "nonsolution" or "constructive evasion" of the problem of who owns what and what ownership means in an inhospitable place like Antarctica.[17] The Antarctic Treaty has no fixed ending date, though after 1991 any country active in Antarctica can request a conference to review the treaty's operation. Many political scientists see Antarctica as a "political laboratory" in which a rather unique solution to a potentially divisive international problem can be tried out for a while. So far, the experiment has worked spectacularly well, and the Antarctic has been looked at as a model for similar treaties on outer space and on the seabed.

Indeed, an Outer Space Treaty was signed in 1967, modeled largely on the very successful Antarctic Treaty. Since no nation had claimed any celestial bodies as territory, there was no need for the calculated ambiguity which the Antarctic Treaty contained. While American machines had landed on the Moon before the Outer Space Treaty was signed, no nation would recognized a territorial claim based on a robot landing, and no claims were in fact made. The Outer Space Treaty is not a final solution for establishing space law, for it was written when space technology was quite primitive and leaves many issues in doubt. For example, this treaty does not demilitarize outer space; it only prevents placing "weapons of mass destruction" in orbit. All kinds of futuristic weapons systems, as long as they were not directed against large populations, could be put in orbit in compliance with the Outer Space Treaty. Still, it's a start.

THE ANTARCTIC: PRELUDE TO SPACE?

It would be nice to suppose that the reason for all of this interest in the Antarctic is purely scientific. Is it possible that 11 wealthy nations are supporting hundreds of scientists to live through the Antarctic winter, and thousands to tramp over the continent in the summer, simply because of the profound scientific possibilities that the Antarctic affords? Antarctic science motivates these scientists to make the personal sacrifices needed to travel to the Antarctic and do science. The United States polar program is bureaucratically located within the National Science Foundation, a government organization devoted to the support of science.

On the face of it, then, it would be possible to make a case that the exploration of the Antarctic is a purely scientific enterprise. This viewpoint, however appealing, is naive; exploration of the Antarctic contains the same curious blend of science, commerce, and national pride which characterized so much of the rest of the Second Great Age of Discovery. There have been many, many occasions in which some commercial possibilities of the Antarctic have appeared. Furthermore, it seems clear that one reason that the American State Department supported the IGY was that it provided a perfect excuse to institute a *de facto* American and Soviet hegemony over the continent without the embarrassing necessity of invoking meaningless territorial claims and stirring up unnecessary conflict.

The commercial potential of the Antarctic has been omnipresent yet elusive. In the 1970s, some scientists were concerned that the krill, the microscopic shrimp which are excellent fish food, might turn out to be excellent people food too. Half a million tons of krill were harvested in 1979–80, largely by Soviet and Japanese trawlers. But krill, when taken from the frigid waters of the Southern Ocean, turns to tasteless mush, making tofu seem like a gourmet's delight. The Antarctic nations recognized krill's critical position in the Antarctic food chain and are conducting a cooperative research program to find out just how much krill can be harvested without wrecking the Antarctic ecosystem.[18]

The large-scale exploitation of other Antarctic resources seems yet more distant. Mineral resources exist, no doubt; coal seams were recognized as early as Scott's abortive polar expedition in 1911. It's an open question as to whether they can be mined economically; the Antarctic is far more forbidding than the Arctic, and the Arctic is tough enough. Another possibility is that freshwater ice, which exists in abundance around the Antarctic as bergs calved from Antarctic glaciers, could be towed to desert regions and provide water for domestic and agricultural use. Many of these ideas have been around for a long time, and it's not clear how soon any of them will become real. In 1988, modifications of the Antarctic Treaty which allow for the possibility of commercial exploitation, and which protect a particular country's investment in mining equipment, was worked out.

However, small-scale commercial activities in the Antarctic are under way already. A number of private companies are taking tourists to Antarctica in airplanes or in boats, and Antarctic scientists are becoming concerned that they may have to spend most of their time showing tourists around when they show up at Antarctic stations, or rescuing them when they are in distress, rather than doing science. Ice cubes from Antarctic glaciers are curios because they contain little bubbles of compressed air which fizz when the cubes melt. At least some people think the exploitation of the Antarctic may come quite soon.

What, then, does the muse of history have to say about the type of space program which the Western world has the political will to support? Some try to draw great hope from the Second Great Age of Discovery. Invoking the name of Columbus as the predecessor of the astronauts is, as we have seen, stretching the facts well past the breaking point. Can you invoke the name of Captain Cook in the way that's usually done: "If you're not sure we ought to explore space, think what the world would be like without the exploits of Captain Cook." "Captain *who*?" would, most likely, be the reaction in much of the world. The real message from the history of the Second Great Age of Discovery is that a program of scientific exploration can be sustained over an extended period, if it has a substantial military or commercial component to it.

Motivations for the Space Program

Where, then, does all of this history leave the four major scenarios depicted in Chapter 1? To recapitulate, the "full space settlement" scenario is the space enthusiasts' dream, where genuine commercial activity can sustain extensive settlements in orbit near the earth and beyond. There is no question that the political will exists to support the "full space settlement" scenario, if it can work, and if its success can be demonstrated fairly early. When the 18 survivors of Magellan's expedition could bring back enough valuable cargo to make this journey profitable, there was then no problem in finding backers for more ventures.

The history of exploration, particularly of the Antarctic, is more closely tied to some of the other scenarios for the future of space exploration. In the "research and tourism" scenario, for example, humans do exist in space, but the reason for their existence is to do science and to simply enjoy the experience of being there. Tourism in the Antarctic has just barely begun, with a few venturesome travel entrepreneurs sending passenger ships to the Antarctic Peninsula in the southern hemisphere summer. The "space science only" scenario is somewhat similar, without the large-scale presence of people. Our use of space would be limited to scientific exploration.

As a space scientist, I would really like to believe that I and my colleagues could cite James Cook as our historical ancestor, even though the likely response among many members of the general public is "Captain who?" Cook was, after all, able to get governmental support for his exploring expeditions. However, claiming such historical ancestry would be grossly overstating the "purity" of the motives of Cook, Humboldt, and the Antarctic explorers. One reason Cook tried to search for the Antarctic was to lay claim to a lush southern continent in the name of the British Crown. When he failed to find such a continent, the scientific returns from the Cook expeditions then appeared to be the main reason for the expedition's existence. However, I doubt that the 18th-century British government would have supported Cook on an expedition whose sole purpose was to discover new species of plants and animals.

Antarctica, again, belies the image of "pure science" which some space enthusiasts have tried to evoke. Even when Antarctica turned out to be a comparatively tiny, ice-bound continent, whaling and sealing interests in its shores have remained to this day. One primary reason that over a dozen countries maintain bases down there is that the possibility of commercial exploitation remains. Doing science in the Antarctic is a way of covering a nation's bets in the name of science.

There is the political will to support a research-oriented space program, provided that there is a significant commercial component to it. My own experience in the contemporary political scene is entirely consistent with this lesson from history. An exploratory program which blends science with commercial applications, in the way that was done in the Second Great Age of Discovery, can survive in the political arena. Research for its own sake simply can't.

POSTSCRIPT: THE ROLE OF NATIONAL PRIDE

But what of national pride? Historian Vernon Van Dyke used the phrase "pride and power" as the title for his book on why we went to the moon. The idea of a red flag with a yellow hammer and sickle standing, unchallenged, in the colorless lunar soil would be an insult to the American concept of preeminence, so we decided to get there first. Could an American President, in the opening years of the 21st century, tell horror stories of the Soviet flag flying in the thin winds of Mars and thus gain support for an American expedition, in the event that prospects for a joint Soviet–American expedition fail? Could national pride alone support a sustained program of space exploration?

I doubt it, for 500 years ago a country tried to sustain an exploration program on the basis of national pride, and this program has vanished into the black hole of history. One hundred years before Columbus, great fleets of huge vessels sailed over the oceans, carrying a powerful nation's flag to distant shores. These ships were not the caravels of Prince Henry the Navigator of Portugal, seeking a new trade route to the spice islands. Nor were they the single-sailed vessels of the Vikings, seeking more plunder.

These fleets were sent by the Chinese empire, led by Admiral Zheng Ho.* Zheng *who*? This man, also known as the Admiral of the Triple Treasure, led a series of expeditions which had the technological capability to discover America, Australia, and Cape Horn. In fact, his expeditions did explore large parts of the Indian Ocean. Simple chauvinism may account for his absence from the history books. A more likely reason for his obscurity is that his expeditions, no matter how successful in their time, were abandoned after he died. I hope that the American space program does not suffer a similar fate, but some of the ingredients which caused the Chinese exploratory effort to self-destruct are also present within NASA.

In 1405, almost a century before Columbus, Zheng set off on one of the most extravagant exploratory missions ever. Flotillas of as many as 300 ships, with up to 37,000 sailors, pushed farther and farther west, going everywhere in the China Sea and the Indian Ocean. The largest ship in these expeditions, the Treasure Ship, was 444 feet long and 180 feet wide, about half the size of a modern supertanker or aircraft carrier.

Why did they sail? Vernon Van Dyke's phrase "pride and power," used to describe the motives for the space program, is an excellent shorthand description of their motives, though the emphasis is on pride. When Daniel Boorstin writes that the Chinese hoped to make their trading partners "into voluntary admirers of the one and only center of civilization," it sounds a great deal like "preeminence," the phrase which has haunted the space program since the 1960s.[19] Apparently the sole motive for this

*Some authors spell his name Cheng Ho, using the traditional, so-called "Wade-Giles" system of transcribing Chinese characters.

extravagant venture was to glorify and solidify Chinese prestige. There was no real thought of commercial gain, scientific exploration or natural history, religion, or military power; the fifth expedition did bring back a couple of giraffes to Beijing.

Whatever commercial trading went on bled the Chinese economy rather than contributing to it. The huge Treasure Ship apparently carried rich gifts, finely crafted, which were given to local leaders in order to impress them. A stone slab on the southeast coast of Sri Lanka, south of India, lists presents: 1000 pieces of gold, 5000 pieces of silver, 100 rolls of silk, a ton and a half of perfumed oil, and some lacquered bronze ornaments. This is Viking plunder in reverse!

What happened? Why is the name of Zheng Ho on the historical junkpile? Zheng Ho, the "Three Jeweled Eunuch," fell victim to some bureaucratic struggles within China which are strikingly reminiscent of contemporary battles between NASA and White House insiders. Chinese eunuchs and the regular bureaucracy battled within the Imperial Palace for power, and Zheng Ho's expeditions existed precisely because he, as a eunuch, had frequent access to Emperor Yung Lo and persuaded him that these expeditions were worthwhile. This emperor died, and his successors were less friendly. In 1435, Emperor Ying Zong took over, and he was persuaded that the eunuchs had entirely too much power. Confucian civil servants argued that money was needed to repair the Great Wall of China and fight off the Mongols rather than send huge treasure ships gallivanting over the Indian Ocean. Historian Daniel Boorstin writes, "Fully equipped with the technology, the intelligence, and the national resources to become discoverers, the Chinese doomed themselves to be the discovered."[20]

It sounds so much like the early days of NASA. President Lyndon Johnson was one of the space program's foremost proponents; he was one of the insiders who persuaded Kennedy to commit America to a moon trip as a symbolic gesture of American pride. After Johnson left office amid the Vietnam protests, no American president has taken a particular interest in space, and NASA's program has never had the preeminent place in the nation's heart that it did in the 1960s. While many noble pronouncements came out of Ronald Reagan's White House, neither the Reagan Administration nor the few Congressional proponents of space exploration have been able to translate these bold pronouncements into concrete political support. In 1984, Reagan tried to evoke Kennedy by committing the nation to build a space station within a decade. By the time the decade was half over, in the late 1980s, the commitment to building the space station was still mushy, with the station program becoming more and more of a political football and its future becoming clouded by the prospects of future budget deficits.

*　　　*　　　*

Space enthusiasts have attempted to use historical precedent to rally popular support. Certainly, any nation which spurned the Columbus expedition, as Portugal did, would live to regret its decision. However, the expeditions of Columbus and Magellan had an immediate commercial return, which is lacking in the space program. Subsequent expeditions like those of James Cook are considerably less famous, not often used to justify the space program. Even they, though, were not just science and exploration, for the commercial and territorial motives were not far away.

Human exploration of the Antarctic has often been cited as a historical precedent for the space program. While it's true that many of the mechanisms for international cooperation and for dealing with the question of territorial claims have been applied directly to space, the motives for Antarctic exploration are not the pure science ones which space enthusiasts often cite as a reason for going into space. Rather, the possible commercial use of the Antarctic has always been in the background.

History, then, does not provide a precedent for a "pure science" program of exploration into the heavens. Of course, scientists have participated in many exploratory ventures, going along on expeditions like those of James Cook. Once a political decision has been made to establish bases in a place like Antarctica, then governments have been quite generous in sending scientists to those bases and supporting scientific expeditions from them. However, the historical message seems clear: if the space program is to be continued in some sustained way, some kind of tangible return must appear in the relatively near future.

CHAPTER 4

Thirty Years Into Space

It has been over three decades since the space age began with the spectacular surprise, the launch of the Soviet Union's Sputnik 1 satellite in late 1957. The American space program sent astronauts to the moon, built a space shuttle, and then faltered when the space shuttle turned out to be a rather expensive and dangerous way of putting hardware into orbit rather than the inexpensive, reliable space truck which it was sold as. The Soviet space program, which stumbled on its way to the moon, plugged along and launched a series of space stations in low earth orbit. The Soviets now have considerably more experience with long-duration space flight than the Americans have.

Thirty years of experience with space activities have set in place a number of attitudes about space which have some rather strong parallels with other historical explorations. Just as any European nation which aspired to leadership in the 18th century had to have a navy that could sail to overseas colonies, all of the major countries today have their own space programs, and a nation which falters from a preeminent position is seen by its own citizens and by others as being in trouble. Space activity is a part of human culture. Space programs, however, are no longer simple, unilateral activities with but a single goal; the conflicts between the military and civilian uses of space, and between sending astronauts or machines into space, persist to this day.

The American space program has been widely criticized for a lack of long-range planning. In the past, it has tended to focus on one megaproject at a time. This approach worked for Apollo, but not for the shuttle. A symptom of the absence of long-range planning is that the two key questions which will govern our use of space in the next century—whether we can obtain the resources needed to support humans in space from space itself, and whether space industrialization can occur—have been far from the center of NASA's attention. The Soviets' long-term plans are less clear, but they have had far more experience with supporting human life in space for long periods of time.

While 21st-century space activities are not imprisoned by the attitudes we hold now, space programs in many countries have developed a certain amount of momentum. Any drastic changes in their direction will risk stalling the effort completely. It is certainly reasonable to look to recent history, as well as to the distant past, to ask what kind of space program is politically reasonable in the contemporary context. The conclusion from recent history is the same as the conclusion from the more distant past: If an extensive program of space settlement is to happen, it must be relatively inexpensive, obtaining needed resources from space itself, and it must be able to generate some sort of commercial revenues.

SPACE EXPLORATION, SPUTNIK, AND CULTURE SHOCK

October 4, 1957: Lyndon Johnson's guests ventured forth into the cool autumn breeze, strolling along the Pedernales river in southern Texas and gazing skyward. They had just heard that the Soviet Union had launched Sputnik 1, 184 pounds of metal which circled the earth every 90 minutes, beeping radio signals down to Americans, Russians, and anyone else who cared to listen. The long-awaited American Vanguard project was still on the launching pad at Cape Canaveral, Florida. The friendly, familiar stars twinkled downward, but amid these familiar signs of night there was something else, a fast-moving point of light reflected off a metallic sphere, easily visible to the naked eye. The sky seemed threatening, but it was not just familiar dark clouds that made Johnson and his friends uneasy. Johnson later wrote that "this evening the sky seemed almost alien."[1]

People who want to hold anniversary celebrations often have difficulty describing just when a particular human enterprise began, but setting the start of the space age is really quite easy. The launch of Sputnik was a sharp salvo in the Cold War which then existed between the two global superpowers. The Russian achievement was a complete surprise to the Western public, for they had ignored, overlooked, or discounted many reports of Soviet plans.[2] The openness of *glasnost*, which might have provided some credible backing to the sparse statements of Soviet intentions, lay 30 years in the future.

Where do we stand now, 30 years later? The space race of the 1960s, the 33-month pause in American shuttle launches which followed the *Challenger* explosion, and the continued stream of Soviet successes have caused many commentators to state that the Soviet Union is "ahead" in space. In many important spheres of space activity, they are indeed ahead, but activity in space is sufficiently complex and multifaceted that it is not possible to say unequivocally that one country is preeminent in space and the other is far behind. The notion of preeminence, stemming from a 1962 speech by President Kennedy, made sense at a time when the single major space activity was landing on the moon, but it is hard to apply in the current context.

Astronaut Sally Ride's 1987 planning document identified several ways of attempting to assess space leadership, and I will outline them here to illustrate how complex the space program is.[3] Any particular type of space activity passes through four stages, in Ride's categorization: First is the *pioneer* stage in which the very first steps in some area of research or exploration take place. Following this is a *complex second* stage in which a pioneering effort broadens, multiple goals are established, and the technology becomes more complicated. Space activities then proceed to the *operational* stage which, with NASA, should not (and now does not) mean operating like a trucking company but means that most activities involve routine applications of well known technologies rather than technological innovation. (Incidentally, in her categorization, Ride does not regard the shuttle as "operational" but places it as a complex second.) In the final stage, a technology becomes *commercially viable*.

Ride also identified a number of different spheres of space activity. Starting from the earth, they are low earth orbit, high earth orbit, the inner solar system (as far as Mars), the outer solar system, and deep space—the universe beyond the solar system. In Table 2, I've added one area to this: the comets and asteroids, primitive bodies which may someday be a key resource in our settlement of the inner solar system. Ride adds "supporting technologies" as another entry in the table; I've simplified this by referring only to launch vehicles.

Any particular space activity can be visualized as filling in a place in a table or two-dimensional matrix, falling into some physical place in outer space and some conceptual place in a stage of exploration. The Apollo 11 lunar landing mission, for example, was clearly a pioneer mission to a location within the inner solar system. By the time we got to Apollo 15, the mission was a "complex second," with a little lunar rover on board allowing the astronauts to ride around the lunar surface.

Table 2. Representative Space Activities in the 1957–1977 Time Frame[a]

Region of Space	Leadership Stage: Pioneer	Complex Second	Operational	Commercial Viability
Deep space	OAO-2			
Outer solar system	PIONEER			
Comets and asteroids				
Inner solar system	VIKING APOLLO 11 Veneras	APOLLO 15		
High earth orbit	SYNCOM		EARLY BIRD	WESTAR
Low earth orbit	Sputnik Vostok 1 MERCURY	GEMINI SKYLAB	Soyuz	
Launch vehicles	SATURN 5			

[a]American ventures are capitalized; Russian are in upper and lower case.

In the 1960s and even in the early 1970s, only a few cells in this table, at the bottom left in the early stages of exploration and close to the earth, were accessible. The pattern of our activities was such that a clear pattern of leadership and, for someone who insists on using the word, American preeminence began to emerge. Table 2 is a simplified version of the similar table in Sally Ride's report.[4]

In low earth orbit, the Soviet Union was clearly the leader in the pioneer stage with its Sputnik earth satellite and Vostok 1, the first vehicle to carry a man into space, in 1961. While the American Mercury program lagged behind the Soviets, Ride places the United States ahead in the complex second phase with its two-person Gemini spacecraft carrying two people and with the Skylab space station. While one could argue that the Soviet Soyuz space stations were comparable in sophistication to Skylab, if not in volume, America's general position of preeminence remains clear. The Syncom, Early Bird, and Westar series of communications satellites in high earth orbit were all American, though by the end of this period the Soviet Union had established its own commercial communications satellite system. Beyond earth orbit, the United States had the Viking landing on Mars and the Pioneer mission to Jupiter, well ahead of the Soviet Union's efforts, consisting of many partially successful Venera missions to Venus, which only succeeded in returning pictures of the surface in 1975. In the deep space area, the American OAO-2 (OAO =

Orbiting Astronomical Observatory) telescope is used, one of several used to survey the universe in the ultraviolet part of the spectrum, a part which does not penetrate the earth's atmosphere and must be observed from space. In the launch vehicle area, the United States clearly had more capability with its Saturn 5 moon rocket, capable of lifting several hundred thousand pounds into earth orbit.

In the 1980s, however, the picture became considerably more complicated as the space programs began to proliferate (Table 3). In launch vehicle technology, the United States lost even its pioneer lead when it phased out the Saturn 5 moon rocket in an incredibly misguided decision to rely entirely on the space shuttle for access to space. The Soviet Union's Energia rocket is perceived by some observers as a symbol of leadership, even though it is basically the same as a Saturn 5. Many of the planetary and space science payloads would not be sitting on the ground if America still had the Saturn 5. While the American shuttle is a new complex second technology, the American penchant for always using more complex technology left us behind in the operational area, where the Soviet Union is so confident in the reliability of its old-fashioned Semyorka rocket that it launches them (without people on board) during snowstorms. Commercially viable rockets include the French–European Ariane and the Chinese Long March rockets; I have not listed any American rockets because an American private launch vehicle industry is not yet established.

In low earth orbit, there are no pioneers because that's been done already. The United States is ahead in the "complex second" area because, as Ride sees it (and I agree), the satellite repair capabilities of the space shuttle represent a new technological area which the Soviet Union has not yet ventured into. The Soviet Union is clearly the leader in operational facilities with its Salyut and Mir space stations. Leadership in the commercially viable area goes to a new contender, the French quasigovernmental Spotimages corporation, which can provide the sharpest pictures of anywhere on earth taken from orbit. When the Chernobyl reactor caught on fire, it was pictures from the SPOT satellite (SPOT = Système Probatoire pour L'Observation de la Terre) which appeared in the newspapers.

In high earth orbit, the United States Global Positioning System, a system of satellites which can allow anyone to determine their own location to within a few hundred feet,[5] is a leading entry, as is NASA's TDRS (Tracking and Data Relay Satellite), a sophisticated way of communicating with its own satellites which alleviates the problem of maintaining tracking centers around the world. NASA satellites beam radio signals to

Table 3. Space Activities from 1977 through 1991[a]

Region of Space	Leadership Stage			
	Pioneer	Complex Second	Operational	Commercial Viability
Deep space	IRAS SPACE TELESCOPE Mir–Kvant Ginga			
Outer solar system	VOYAGER	GALILEO		
Comets and asteroids	Giotto Vega Sakigake			
Inner solar system		Phobos MAGELLAN	Venera	
High earth orbit		GLOBAL POSITIONING SYSTEM	TRDS	COMMUNICATION SATELLITES
Low earth orbit			Mir Salyut	Spot
Launch vehicles	SATURN 5 (DISCARDED)	SHUTTLE Energia	Semyorka	Ariane Long March

[a]Capitalized ventures are American; upper and lower case are Soviet; underlining denotes other nations' programs.

the TDRS, which then communicates with civilian or military control centers. While many of the world's communications satellites are run by the international consortium INTELSAT, I have followed Ride in giving the leadership nod to America because American companies build most of these satellites.

It is in the exploration of the inner solar system that the greatest change has taken place. American leadership has given way to the Soviet Union, where the Phobos mission to one of Mars's satellites and the Venera radar imagers of Venus come way ahead of the American Magellan mission. Magellan will produce radar pictures of Venus which won't be much sharper than those from Veneras 15 and 16, and its launch has been long delayed because of space shuttle problems.

New countries have emerged in leadership positions for exploring the comets and asteroids. The Soviet Union, ESA (the European Space Agency), and the Japanese all sent missions to Halley's Comet. NASA's redirection of another satellite to Comet Giacobini-Zinner might have been scientifically as valuable as the Halley flotilla, but leadership also includes visibility, and clearly a mission to Comet Giacobini-Zinner is not as visible as a mission to Comet Halley.

Beyond the inner solar system, the United States is still clearly in the lead, though not in the position of undisputed preeminence that it was in during the 1970s. The Voyager mission which visited Jupiter, Saturn, Uranus, and Neptune was a remarkably long-lived spacecraft and brought back spectacular views of the varied moons in the outer solar system. Galileo, which will orbit Jupiter in 1994, is an excellent complex second. The IRAS (Infrared Astronomy Satellite) which explored the universe in the infrared and discovered dusty disks, protoplanetary clouds about to form planets, around nearby stars was a cooperative venture between the United States, the Netherlands, and Britain. The 90-inch diameter Hubble Space Telescope dwarfs the Soviet telescopes aboard the Kvant astrophysics module attached to the Soviet Mir space station. The Japanese entered the field of deep space by launching their Ginga X-ray observatory in the mid-1980s.

It's a complex picture, even though I've left out 90% of the missions which have flown or will fly in the period in question. I find it hard to see a clear leadership pattern in the 1980s and 1990s. The interest in a Mars trip which emerged in the late 1980s focused attention on two key areas where the Soviet Union is indeed ahead: in the exploration of Mars itself, and in understanding the effects of long-duration spaceflight on humans. The Soviets' ability to continue to launch missions for a period of nearly three years when the American shuttle was grounded highlighted a temporary American inferiority, since it is clearly impossible to have a space program if you can't launch heavy hardware into space.

MILITARY AND CIVILIAN SPACE PROGRAMS

The preceding discussion of Tables 2 and 3 treated the civilian space programs of each country, but made no mention of each country's military programs. Many people would like to see civilian and military space programs separated, and view any hint of military activity as somehow tainting a purely civilian NASA. However, the military and civilian space programs have been intertwined for a long time, and indeed it was the military's space program which got America into the space business quickly after Sputnik.

The American Reaction to Sputnik

Many Americans, including then President Dwight Eisenhower, were impressed but unalarmed by Sputnik. The media, however, panicked. Stories replete with banner headlines claimed that a nation which launched the first satellite could also launch H-bombs onto our doorstep. H-bombs weigh a lot more than the Sputnik satellite, but newspaper editors considered that as a minor detail. The triumph of Sputnik was seen as heralding anything ranging from increased Soviet influence in the Third World to the military collapse of the Western alliance.

One group of new Americans who agonized over Sputnik was a group of German-born rocket scientists in Huntsville, Alabama. Nazi Germany had developed the V-2 missile in World War II; these missiles landed on London from launching sites in Peenemunde, Germany, east of Denmark. Wernher von Braun and his associates deliberately decided to surrender to the Americans rather than the Russians. Nazi Germany was a rather confused place in April 1945, just before the end of World War II; the German rocketeers obeyed the orders to move southwest, and ignored the conflicting orders to stay in Peenemunde (where they would have surrendered to advancing Soviet armies). On April 30, 1945, Magnus Von Braun, Wernher's English-speaking brother, sought out the Americans, surrendered to Private Fred P. Schneiker in the small town of Reutte, and told him that 150 German rocket engineers were just waiting to be captured.[6] Surprisingly, the rocket assembly lines in Nordhausen were intact too. Eventually people, blueprints, and rockets were all moved to the United States, first to Fort Bliss, Texas, and eventually to Huntsville, Alabama. The German rocket team became part of the Army's intercontinental ballistic missile effort.

In 1956, though, it seemed that this rocket team was to be shut off from the American satellite-launching effort. A high-level advisory group chose the Naval Research Laboratory as America's entry into the satellite program, turning its back on von Braun's rocket expertise. Apparently, the political desire to separate the civilian satellite program from the military missile program played a considerable role in the choice.[7] Von Braun could continue to work on missiles, but was not allowed to work on satellites. This separation of "military" and "civilian" space programs seems, in retrospect, particularly artificial, since America developed the "civilian" space program at least partly to establish its right to fly picture-taking satellites over another Russian territory.[8]

The Army group did indeed continue developing missiles, fully aware that they could replace the dummy warhead on top of their workhorse Jupiter C rocket with a tiny rocket which could propel a hunk of metal into orbit. One reason they may have lost the satellite project to the Naval Research Lab's Vanguard project is that they had not put much thought into what the hunk of metal would do once you got it into orbit. The Navy had much better contacts in the scientific community, and the scientific instruments on their satellite were seen as being superior.

Well before Sputnik, a number of American scientists were developing machinery and electronics which could explore the earth's upper atmosphere, immediate surroundings, and the universe beyond if they were launched on an orbiting satellite. In 1956, 25 scientific groups proposed different scientific instruments which could be carried on an American satellite. Four of those were selected as priority A, candidates for instruments on the first few vehicles. Professor James Van Allen of the University of Iowa, leader of one of the four selected teams, kept in touch with both the Navy and Army rocket programs, and was well aware of the Army's progress and the Navy's troubles. A year before Sputnik, Van Allen decided that he would design his instrument so that it could be launched either as part of the Naval Research Lab's Vanguard program or on top of Von Braun's well-tested rocket.

The result of the congressional and media outcry was that Von Braun's rocket team was unleashed at the end of October 1957 and given the go-ahead to launch a satellite. Von Braun boldly predicted that he could launch a satellite in 90 days, and hard work on the part of Von Braun's team produced success. Van Allen and his associates also worked hard to get their payload ready. In December, Project Vanguard failed ignominiously; the rocket lifted itself a majestic four feet above the ground and slumped backward in a cloud of foul, black smoke. Von Braun met his deadline and launched Explorer 1 on January 31, 1958.[9]

The Explorer story provides a lesson from history: the space program is multifaceted, consisting of a mixture of "civilian" and "military" technologies and goals. Because Project Vanguard was supposed to be "civilian" (even though it was run by the Naval Research Laboratory, a military-affiliated organizaton), it could not draw on the decades of expertise represented by Von Braun's rocket team. The Naval Research Laboratory had to reinvent the wheel, building its own booster, the so-called Viking rocket. While it may be more convenient for Western countries to have separate military and civilian budgets and launch facilities, these labels cannot change the close relationship between the civilian and military uses of space. It seems to me that when you make future plans for the space program, you have to accept this close relationship as an established fact.

Conflict in the 1980s: Star Wars and the Shuttle Schedule

The apparent conflict between military and space programs in the 1950s subsided through the 1960s and 1970s, when NASA went its way and the military space programs went theirs. These decades showed that it was quite possible for a civilian and a military space program to coexist side by side. During this time, there was no competition for resources between the military and civilian space programs, for each had its own budget to buy rockets from whatever American company could build the particular rocket which it wanted. The budgetary climate of the time meant that there was never a time in Congress when an explicit tradeoff between defense spending and spending on other things had to be made.

In the 1980s, though, the conflict between military and civilian space programs erupted again as a result of two developments which forced the military and civilian space programs to compete for different types of resources. The decision to rely on the shuttle as the only launch vehicle resulted in a competition between military and civilian missions for scarce slots on a crowded shuttle schedule. The pressure to launch mission after mission is widely cited as one of the causes of the *Challenger* accident. When *Challenger* did explode the situation became even worse, for shuttle launches became scarcer yet. The military was down to one aging spy satellite and was anxious to get replacements into orbit.

Another 1980s development brought the military and civilian space programs into competition for public attention and for money: Star Wars. There are many misconceptions about what Reagan's Strategic Defense Initiative (SDI) is and is not. Research on defensive schemes, and even on the futuristic particle-beam technologies, had been going on well before Reagan's 1983 Star Wars speech. The speech and its aftermath generated a strong movement toward an accelerated research program. Many outside the Star Wars program still harbor the illusion that a defensive shield which will ward off a massive enemy attack is just around the corner; scientists within the Star Wars program as well as outsiders repeatedly emphasize that such a shield, if it can ever be implemented, is a long way away. In the waning years of the Reagan Administration, SDI's popularity within the Pentagon faded as many realized that it was soaking up many billions of dollars and starving other weapons development programs.

However, if even a limited SDI system were to be deployed in the 1990s, the pressure on the shuttle launch schedule would be tremendous, because of the large number of shuttle flights needed to place the necessary hardware in orbit. Between SDI and the space station, there would be so much demand for shuttle launches that anyone else seeking to launch a payload would have a hard time slipping into the shuttle launch schedule. In addition, preoccupation with the Federal deficit in the late 1980s resulted in Congress's finally coming to grips with the question of tradeoffs, and so NASA and SDI are in a very real sense competing for the same pot of money.

Apart from SDI, though, the conflict between military and civilian space programs shows some signs of abating in the short run. The Air Force, the military service with the prime interest in space, had won a long bureaucratic battle with NASA and began to buy some Titan rockets in 1984, a couple of years before *Challenger*. The Air Force has shown a great reluctance to launch military payloads on the shuttle, apparently partly because it is reluctant to cede control of a military program to another agency (NASA) and partly because there are security concerns about launching military payloads from the NASA's Kennedy Space Center, a civilian installation with limited ability to keep people away from a secret military payload. As long as something like SDI doesn't emerge in the early 1990s, pressure on the shuttle launch schedule won't come from military-civilian competition, but rather from the considerable requirements of the U.S. space station.

THE ASTRONAUT'S ROLE

The special role of astronauts in the space program has turned out to be a double-edged sword. They can make us feel exceptionally proud and patriotic when they succeed, but when tragedy befalls the astronauts, our anguish is magnified. America's first taste of tragedy occurred in January 1967, when Roger Chaffee, Edward White, and Virgil Grissom were incinerated in an Apollo capsule when an electrical fire grew immediately to inferno intensity in the capsule's pure oxygen atmosphere. The ensuing congressional investigation, paroxysms of national self-doubt, and necessary changes in machinery set the Apollo program back by at least a year.

More recently, of course, is the *Challenger* disaster. The O-rings, which were supposed to flex and seal the joints in a solid booster rocket, didn't flex and seal in the freezing temperatures of January 1986. At 73 seconds after launch, exhaust gases escaping from the solid rocket ignited the huge orange tank carrying liquid hydrogen and oxygen, transforming it into a giant orange bomb which incinerated the space shuttle. Miraculously, at least two of the astronauts survived the initial fireball, only to perish sometime later as the spacecraft plummeted into the Atlantic Ocean.

Challenger was a much more serious setback to NASA's plans than the Apollo fire was. NASA had ambitiously scheduled 15 shuttle launches for 1986, and had only one successful mission, one that flew a few weeks before *Challenger*. The 15 launches planned for 1987 were all scrubbed, and the launch rate will shrink considerably into the indefinite future. Most observers believe that it won't be until the mid-1990s that the effects of this disaster wear off.

Public reaction to *Challenger* indicated a really serious nonfinancial cost to the space shuttle program. Because astronauts are heroes whom many Americans identify with, there is tremendous pressure to make the space shuttle "safe." It is often difficult for people to realize that nothing is safe, that anything which we do in our lives carries some risk, and that people get killed in car crashes as well as in *Challenger* explosions. We have accepted the reality of automobile accidents in a way that has not applied to the risks of space travel.

There were a number of special circumstances surrounding the *Challenger* disaster which made it impossible for NASA to follow some people's suggestions and continue with shuttle launches as long as the weather was warm enough to pose no danger to the O-rings. Briefly, the *Challenger* exploded because the O-rings, huge rubber washers which were supposed to seal the joints in the solid rocket boosters, lost their flexibility in cold weather and didn't seal the joints properly. *Challenger* exposed weaknesses within NASA's management structure as well as bad O-rings and a bad joint design. I will not recount in detail the decision-making process which led to the launch of the Space Shuttle *Challenger* in January 1986. Briefly, an engineer's "no go" recommendation turned into a "go" as the message went up the decision-making pipeline within NASA and its contractor Morton Thiokol.[10] When these events became public, there was great disillusionment with an agency which had been admired more than any other in government. The Pentagon buys $500 hammers and $3000 toilet seats. Social Security and the Veterans Administration call live people dead and send dead people Social Security and veteran's benefit checks. But NASA, before 1986, brought us the stars. After 1986, it became clear that they could screw up just like anybody else.

The remarkably successful flight of *Discovery* in late 1988 put NASA on the road to recovery. A greatly redesigned space shuttle performed flawlessly. Because the shuttle changes didn't make it look different on the outside, it's not easy to appreciate that the space shuttle which took off in the fall of 1988 was a rather different flying machine from the one which blew up in January 1986. To its credit, NASA used the nearly three-year period that it took to redesign the joints in the solid rocket boosters, test them, and test them again to make a number of other changes in the shuttle. An escape system was added, even though the escape system can only work when the shuttle is gliding in a controlled manner and certainly wouldn't have helped the astronauts escape the *Challenger* disaster. Whenever engineers make that many changes in a spacecraft, there's a concern that these changes have affected some other major subsystem in the craft which will create problems. No such problems emerged on *Discovery's* first flight.

Yet NASA still faces some difficulties. It is still true that many shuttle flights are being used to do something which could easily be done by expendable rockets—take a piece of machinery, cart it up into space, and place it in orbit. The shuttle's unique capability of taking people up into space and letting them do experiments and fix hardware when they are up there is only critical to a few shuttle missions. The shuttle launch schedule is still rather crowded, and pressures to keep to the schedule may become rather large.

Furthermore, there is still some question as to whether NASA has a real plan for the future. At a time when the Soviet ambitions to explore Mars have crystallized and been made public, NASA's plans beyond the space station seem too vague to many observers. I believe that part of the trouble is that NASA's perspectives on planning are rooted in its origins in the 1960s, when a single-minded focus on one megaproject like Apollo seemed to be the best way to do business. But now that the space program is multifaceted, a different approach to long-range planning seems called for. It's possible that the Ride report represents just such a change.

SPACE STATION AND NASA'S LONG-RANGE PLANS

Many of my friends yearn for a return to the days of the 1960s. Then, scientists' dreams about going to the moon, articulated in the late 1950s, suddenly and effortlessly became reality when Kennedy gave a speech and everybody lined up behind him. A lot of people hoped that Reagan's similar challenge to NASA in 1984 to build a space station within a decade would generate the same kind of public support. While it seems likely, at this writing, that the space station will eventually be built, it has not had the smooth ride that Apollo had, and even four years after the initial commitment there is some doubt that the necessary money can be found.

Megaprojects and Top-Down Decision Making

Space historian John Logsdon has been telling us for years that the decision to go to the moon was an exceptional decision indeed.[11] While it might seem that understanding the details of Washington decision making is far from the space program, the space community needs to appreciate the limitations of political realism in order to help NASA create a space program which can sustain its momentum for an extended period of time. People in the space community have had a tendency to wait for a presidential speech and, once a speech is made, assume that such a speech will automatically result in political support.

Logsdon's primary thesis is that the decision-making process in government is, in general, a "bottom-up" one. Someone with an idea has to gradually build a base of support among bureaucrats in the relevant agency (such as NASA or the Defense Department), in Congress, and in the White House. A government worker will get some money from somewhere and start work, in a small way. Make enough progress, and this project may get mentioned in a sentence or two in presidential speeches. Get a few senators on board, and they will introduce bills, sometimes just as a way of getting attention rather than as a serious proposal. This slow, gradual process can sometimes lead to a major accomplishment. The National Science Foundation's (NSF) Supercomputer Initiative, a project in which the NSF paid for a half dozen supercomputers and made them available to academic researchers, started in exactly this way.[12]

The Apollo decision, in contrast, was a "top-down" one. The prime movers were President Kennedy and Vice President Johnson. The space advocates responded to a presidential challenge, rather than persuading the administration to do something new.[13] It was Kennedy who asked, "What can we do to beat the Russians?" Under normal governmental circumstances, NASA would be asking Kennedy to endorse a lunar project, and pointing out the benefits from doing so. This type of decision making is rather exceptional.

The congressional and public response to Kennedy's challenge was also very unusual. The unique circumstances also led to remarkably smooth sledding for the space program in Congress, at least in the early years. NASA could scarcely spend money fast enough. In this sense, the early years of NASA were quite different from the early years of almost any other program, where each increase is usually a matter of some percentage, not a doubling or tripling (NASA's budget tripled between 1960 and 1963).

The heritage of Apollo has caused NASA to focus on two more megaprojects, hoping that public support for these will carry the rest of the space program on in the same way that the first planetary exploration spacecraft and the first ventures in many areas of space science were carried along on the coattails of Apollo. The space shuttle is the second of NASA's megaprojects, and the space station, scheduled for construction in the 1990s, is a third.

Neither shuttle nor space station has been able to generate the kind of enthusiasm that the nation had for Project Apollo. The early 1960s were a special kind of time, and Kennedy started Apollo when he seemed to have a magic touch within Washington. After he was assassinated, the new president, Lyndon Johnson, became the only president we have had during the space age who took a particular interest in our space accomplishments.

Appealing to different people in the White House isn't the whole story. The Apollo project was focused on some definite goal, landing on the moon, which many people could easily identify with. A moon landing fits in very nicely with other natural milestones of exploration like climbing Mount Everest, reaching the North and South Poles, and discovering new continents. The shuttle is simply a different way of accessing space, and the space station is simply an intricate machine in earth orbit; neither has the natural appeal of Apollo. People want (and often even need) to achieve new and greater goals rather than simply reaching updated old ones.

Short-Run Plans

What, then, should NASA do in the immediate future, looking toward the 21st century? One of the most important consequences of Sally Ride's planning exercise was a very modest item of only $100 million in NASA's budget, about 1% of the total, called Project Pathfinder. Ride's report identified four possible initiatives whereby the United States could regain leadership in important areas of space activity: a mission to planet earth, renewed exploration of the solar system, a lunar base, and a mission to Mars. The lunar base and Mars missions would require some new technology, and such long-range projects were given little or no resources during the shuttle development era, when every penny that NASA had was cannibalized in order to get the shuttle flying. Project Pathfinder was an effort to begin development of some of this technology.

My own perspective in this book is a bit longer-term than Ride's, and two questions serve to frame the four basic scenarios illustrated in Chapter 1. The first of these questions is very closely related to Project Pathfinder. In Chapter 1, I framed this question as follows: Can extraterrestrial resources be used to support humans in space? The answer to this question is complex and will be explored in the next several chapters of this book. Part of the answer is determining just what it takes to support people in space, and using a nine-month trip to Mars as a focus for such a question is a very reasonable way of asking it.

This first question suggests some additional short range projects for NASA which could help answer just what our future in the 21st century will be. One of the major resources that we need to support people in space is water, and a critical question is whether the moon, the satellites of Mars, or the asteroids which orbit in the inner solar system are possible sources of such water. Many of these bodies are very poorly explored, and a renewed effort to send space probes to them could be considered valuable in its own right, part of one of the solar system exploration initiatives which Ride mentions as a way of regaining American leadership.

The second question which frames possible 21st century space activities is whether any commercial activity going beyond communications satellites will happen, and whether such commercial activity will require a human presence in space. While NASA has tried to bring a number of different companies into the space program, limitations of the shuttle program and the *Challenger* hiatus have discouraged a number of potential partners.

The need to further explore the challenge of supporting humans in space for extended periods and possible commercial activities in space suggests that NASA's proposed investment in a space station is indeed worthwhile. Indeed, these questions give some focus to the space station program which it may lack currently. They are interesting in their own right as well.

* * *

Thirty years of space exploration have left the world's major spacefaring nations in a rather complex situation. The broad nature of our activities in space prevents a clear definition of who is "ahead" or "behind" in space, though the Soviet Union is clearly now in a much stronger position than it was in the mid-1970s after the Apollo moon landing. The complexity of various countries' space programs has created a number of tensions within these programs which are most visible in the American one: military vs. civilian, manned vs. unmanned, and megaproject vs. progress on a broad front are the major ones.

As we look toward the next century in space, NASA's focus on large projects like the shuttle and the space station has left little room for attention to some of the longer-reaching questions that will determine our future in space. Is space industrialization possible? What does it take to support human life in space, and can some of these resources be obtained from space itself? These questions will serve to focus future chapters.

PART II

Living in Space in the 21st Century

Many space enthusiasts fervently believe that the human destiny is to settle the inner solar system. For some, this belief is so strong that it has become an article of faith. Others, most particularly some outspoken space scientists, think that space settlement is a pipe dream which will never come true. The argument about space settlement is not just about imponderable futures; our next moves depend on whether space settlement is likely or not.

In order to assess the likelihood of space settlement, a decision maker needs to know just what kind of resource commitment is needed in order to sustain human life in space. What does it take to establish an inhabitable space station, a lunar base, a colony on Mars, or even a free-floating space colony? Humans have now had a great deal of experience in living in space, and we are now in a position to lay out in some detail just what colonies of various sizes in various places would need. The supply-side aspect of space settlement, then, is the focus for Part II.

CHAPTER 5

The Joys of Space Life

Human beings are the only species of living organisms who have extended their range to include the entire earth. With proper clothing and living facilities, we can survive through the cold of an Antarctic winter, where temperatures range to 90 degrees below zero Fahrenheit, as well as in tropical regions where neither the temperature (again in Fahrenheit) nor the humidity drops below 90 for very long. It might seem reasonable to simply extrapolate our historical experience and postulate that a species which can survive from 90 above to 90 below can also survive in a place with no air (and so the concept of temperature doesn't even make any sense). Besides being naive, such an approach would be wrong, at least partly because one aspect of living in space has never been experienced by humans living on the earth: the experience of zero gravity.

In the early days of the space program, no one knew how humans would react to the absence of gravity. Planners of the Mercury mission assumed the worst, and put the astronauts in a tin can which they couldn't control at all. The designers didn't even want to put in a window at first, not realizing that one of the most exhilarating aspects of space travel is looking down at the earth. The astronaut was a mere test object.

Now we know better. There are a number of rather peculiar things which happen to astronauts when they are in space, but none is fatal and none prevents them from working, eating, sleeping, going to the bathroom, and simply enjoying being alive. (I don't think anybody's tried making love in space yet, and when the "experiment" is first performed I doubt that the rest of us will hear about it.) There are some potential long-term hazards of staying in space for extended periods of time: bone thinning, radiation, and the psychological effects of extended confinement. There are ways around all of these problems, though they may cost a lot.

WEIGHTLESSNESS

Like most writers on the space program, I've never been in space, and I can't report on the joys of space life firsthand. The most unusual experience is the absence of gravity in space. Astronauts within the space shuttle fall freely around the earth in the same way that the shuttle itself does, and so they float around the cabin. To me, the most vivid way of sharing the experience of zero gravity has been seeing the film "The Dream Is Alive" in the large-screen IMAX format. Astronauts simply float through the air. Life seems almost effortless. Astronaut Sally Ride's book, *To Space And Back*, echoes the combined sensation of elation and mystery:

> The best part of being in space is being weightless. It feels wonderful to be able to float without effort; to slither up, down, and around the inside of the shuttle just like a seal; to be upside down as often as I'm right side up and have it make no difference.[1]

The same forces which act on the spacecraft act in the same way on everything inside the spacecraft: astronauts, cups of coffee, books, checklists, tools, and so forth. On the earth, when you let go of this book, the forces which your muscles exerted on the book (which opposed the force of gravity) no longer act on it and the book falls. In space, the book is orbiting around the earth in the same way that you are, and you don't need to use any force to keep the book in front of your eyes.

For millennia, the human body has evolved in an environment where muscles and bones need to oppose the force of gravity. The heart has to work harder pumping blood uphill to your head than it has to work pumping blood sideways to your arms. We have evolved powerful leg muscles in order to walk and get around. When a human being goes into space, the body reacts rather differently. Sometimes these differences make life more pleasant; more often, they create minor physiological difficulties.

Sleeping

Sleeping in space is rather easy. The only difficulty, in the space shuttle, is that the lights are always on; many but not all astronauts donned sleep masks to keep the light out. (Sally Ride reports that the astronauts called these "Lone Ranger masks" because they are black.) The only additional piece of sleeping equipment needed was some kind of restraint to keep the sleeping astronaut from drifting into the middle of the spacecraft and bumping into something or someone. Pillows or expensive mattresses? Forget them. Pictures of astronauts asleep in space make them look almost angelic, as the illustration shows. Astronaut Joe Allen explains:

> Sleeping is perhaps the one daily function that demands *less* design, planning, or deliberation in space than it does on the earth. The eternal fall or orbit provides genuine "flotation support" that is unmatched by any mattress, and the weightless environment demands few special sleeping accoutrements.[2]

There are some drawbacks to the ease of sleeping in orbit. It's always embarrassing to fall asleep when you're supposed to be working. On earth, there are warning systems. Suppose that you are listening to some boring presentation in a meeting, have worked hard all day, and start to slowly, slowly fade away. At the moment that you fall asleep, your neck muscles no longer hold your head in an upright position, and you nod off, with your chin slumping to your chest. At this instant, the nerves on the back of your neck detect the sudden (and sometimes painful) stretching of your neck muscles and snap you back to attention. (At least it usually works. I have to admit that my body has managed to ignore that warning mechanism on occasion, and I have managed to fall asleep in meetings a few times.)

In the weightless environment, though, your head won't fall over, snapping your neck muscles and waking you up. When you fall asleep in space, nothing changes. Your head stays in the same position, and the tools and checklists that you are holding don't fall anywhere because they don't fall. There is no warning clunk as a clipboard falls out of an exhausted astronaut's hand. Awakening will only come when Mission Control starts yelling at the exhausted, nonfunctioning astronaut. Joe Allen reports that the astronauts adopted a buddy system, where the astronauts would try to keep each other awake and Mission Control off their backs.

Minor Body Changes

For some of us at least, there is a significant physiological benefit of being in space. Because there is no stress on the backbone, people add one or two inches in height because the disks, those spongy masses which sit between the bones in your vertebra, aren't compacted by the force of

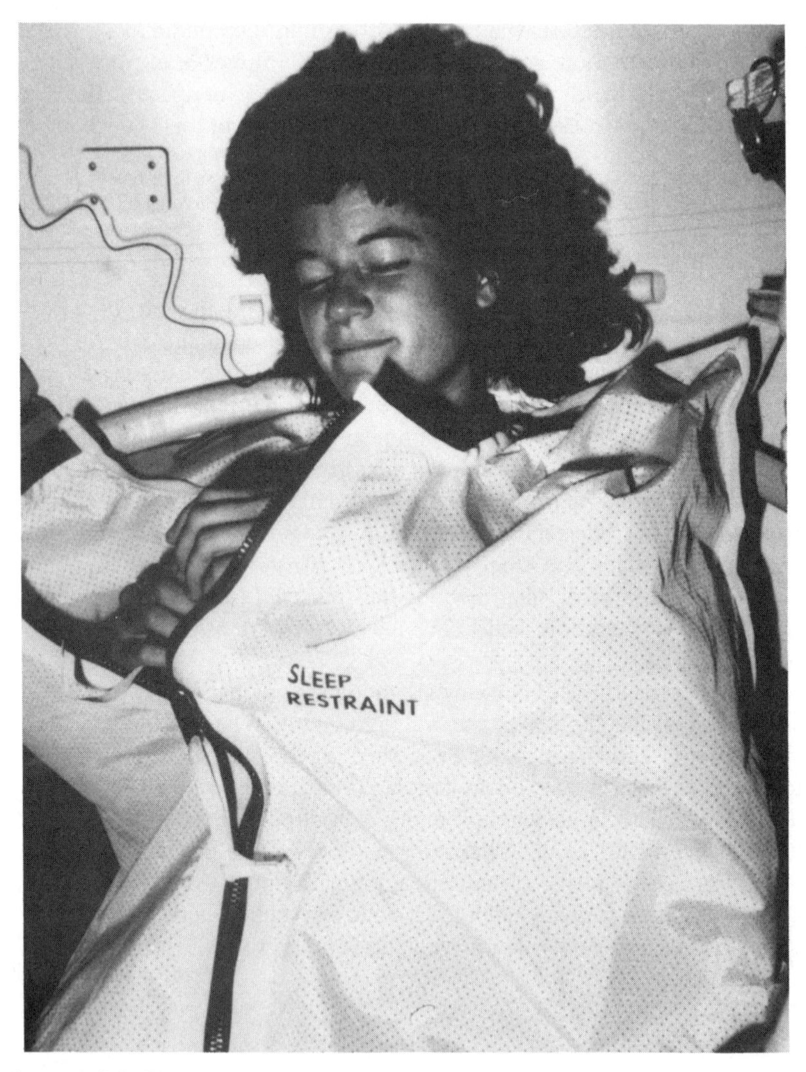

Astronaut Sally Ride was photographed at her sleep station on board the seventh shuttle flight, mission STS-7. The sleep restraint prevents her from drifting about the cabin. (NASA)

gravity. William Pogue reports that his waist measurement shrunk by about 3 inches because his body's internal organs didn't slump downwards. With contemporary American standards, gaining an inch of height and losing three inches around the waist is a real boon. (The effects don't last; when astronauts return to earth, the height is lost and, assuming no permanent weight loss in orbit, the waist measurement is regained.)

The same effects which shrink your waist measurement by a few inches also alter the appearance of people's faces. Fluid, no longer pulled toward your chin by gravity, tends to accumulate around your eyes, making your face look puffy. Psychologists have wondered how this affects nonverbal communication. The familiar smiles which we are so used to will look different in space. Astronauts have not made much of the changed facial expressions, perhaps because people can get used to them in time.

Working

Almost all of us, in our work, pick up things and put them down again. Office workers pick up pieces of paper, books, paper clips, and mail. Plumbers and electricians pick up tools and parts. In space, astronauts use books, fix things, and do experiments, but these activities are a bit different from the way that they work on the ground. It's not possible to put something down; if you let go of it, it either stays where it is, if you let go gently, or it floats away. Spacecraft are festooned with strips of Velcro over every accessible interior surface. Velcro takes the place of gravity in keeping the interior of the spacecraft workably neat, provided that the astronauts remember to use it.

For the most part, the lack of gravity is merely an inconvenience in the first few days of a person's work tour in space. Astronaut William Pogue mentions the example of a new astronaut who picks up a screwdriver and starts to unscrew something. He has forgotten to brace his legs against something and his body starts turning, with the screw remaining frustratingly fixed in its position.[3]

Simply moving in space can present difficulties. Joe Allen again:

> During the first days in space, the act of simply moving from *here* to *there* looks so easy, yet is so challenging. The veteran of zero gravity moves effortlessly and with total control, pushing off from one location and arriving at his destination across the flight deck, his body in proper position to insert his feet into Velcro toe loops and to grasp simultaneously the convenient handhold, all without missing a beat in his tight work schedule. In contrast, the rookies sail across the same path, usually too fast, trying to suppress the instinct to glide headfirst and with vague swimming

> motions. They stop by bumping into the far wall in precisely the wrong position to reach either the toe loops or the handholds. In their attempts to recover before rebounding to the starting point, they twist around too rapidly, knocking loose cameras, film magazines, food packages, and checklists, all previously Velcroed to the wall and now careening about the cabin in different directions.[4]

Usually all these problems subside after a few days in space; astronauts become adapted to these conditions and aren't bothered by them thereafter. But even veterans can sometimes run into difficulties. William Pogue reports that on one Skylab mission, he turned around too quickly to snap a picture of the earth, and his glasses flew off. On earth, when something like that happens, I look at the floor. When I fall asleep at night without taking my glasses off, gravity ensures that there are only a few places that they can be, usually on the floor beside or under the bed. In space, there is no "floor" to look on. Worse yet, because lost objects float around rather than staying put, you can't count on their not being somewhere because you looked there already.

In Skylab there was no gravity to pull Pogue's glasses to some predictable place. He heard them bouncing around the experiment compartment he was in, but then couldn't find them. He did have some spares, but like most people's spare eyeglasses they weren't as satisfactory. Three days later, one of the other astronauts on Skylab found them floating near the ceiling of his bunk.[5]

Inside the space shuttle, rookie difficulties with zero gravity are minor inconveniences, but the problems are potentially more serious when astronauts work outside the space shuttle on spacewalks, more accurately called EVAs (extravehicular activities). In several recent shuttle flights, astronauts have had to retrieve faulty satellites and either fix them or bring them back to earth. Wrestling these satellites back into the cargo bay and fastening them down was no mean trick, since a 1200-pound communications satellite still has inertia in space, even though it has no weight. The astronaut's jet-powered backpacks (of course there's an acronym: aficionados call them MMUs for manned maneuvering units) were powerful enough to stop these satellites' rotation, as long as they were used gingerly. In addition, a long arm, remotely manipulated by an astronaut inside the shuttle, can help the astronauts move around and move satellites around. An astronaut on the end of the arm is shown in the illustration.

Astronauts make some attempts to practice doing all of these tasks in something like a zero-gravity environment. Much preflight practice, especially for EVAs, takes place in a giant swimming pool. Astronauts can be buoyed up so that, effectively, they weigh nothing; pushing one's arms

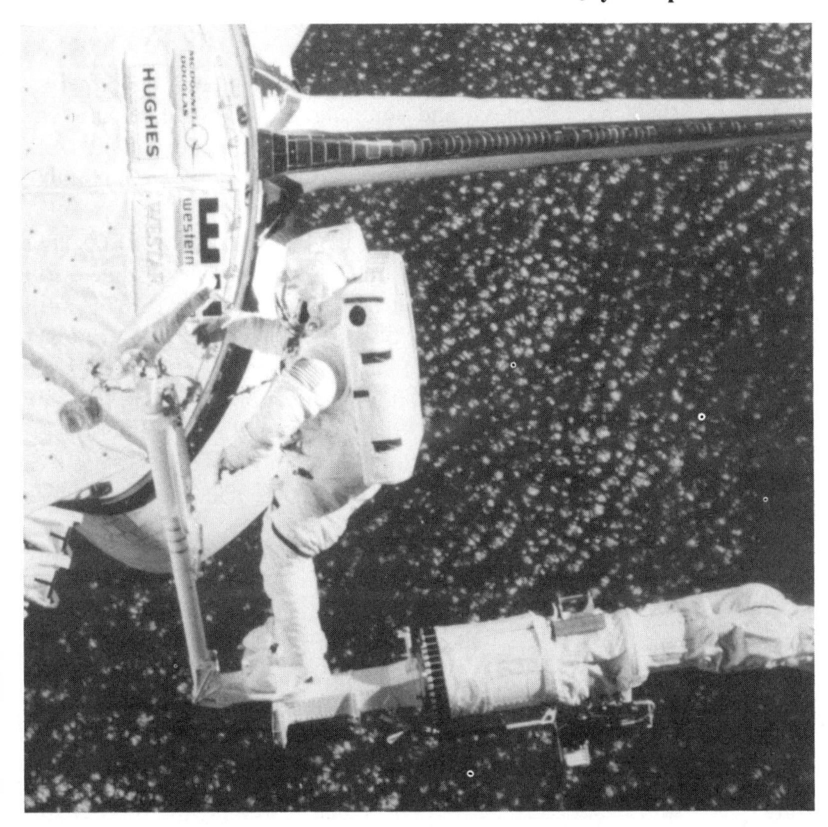

Astronaut Bruce McCandless is perched on the end of the remote manipulator arm, testing the idea that foot restraints on the end of the arm allow another astronaut inside the shuttle to move McCandless around the cargo bay, using the arm as a "cherry-picker." (NASA)

through water is somewhat like moving around in a space suit. In addition, the successful retrieval of two satellites in November 1984 demonstrated that considerable flexibility in the mission is called for, since many of the preplanned schemes for actually grabbing the satellites didn't work.

Space Sickness

About half of the astronauts experience space sickness when they first encounter weightlessness. At one time it was thought to be similar to seasickness, or carsickness, or airsickness. The symptoms can be similar. Space sickness was once delicately called "space adaptation syndrome" by NASA physicians, but more recently they have admitted what it is and called it "space motion sickness" or SMS.[6] Symptoms are lethargy, sleepiness, headache, dizziness, and vomiting. However, there's no way of predicting who will be spacesick; people who are chronically seasick may do just fine in space, while people with iron stomachs on the ground can sometimes have all kinds of trouble when in orbit.

Because space sickness can make it hard for astronauts to do their jobs in important early stages of a mission, NASA has gone to some effort to determine what causes it and what can be done about it. The most famous test subject was United States Senator Jake Garn. The story making the rounds at meetings is that astronauts are rated on an informal scale from Garn 1 to Garn 10, depending on how susceptible they are to being sick. Garn himself breaks the scale, rating a Garn 13. More seriously, NASA's studies have shown that drugs seem to have some positive effects, but the ultimate causes of space sickness remain elusive. It does go away after a few days.[7]

Eating

Here as in other aspects of space life there has been a progression from the ultraconservative, monastic early days to the comparatively civilized conditions aboard the space shuttle. Designers of the Mercury program, fearful of disintegrating bread where fragments would float into and disrupt important electrical circuitry, forced the astronauts to eat pureed stuff out of tubes, basically baby food. To save weight, the food is still dehydrated, since the space shuttle's fuel cells are a source of pure water to rehydrate food.

Just about anything that you can eat on earth could, in principle at least, find its way to the space shuttle menu, though in dehydrated form. A few foods are still off-limits, primarily because foods which aren't sticky will tend to float off of a food tray and wander around the cabin. An astronaut who wanted to eat peas would have to tolerate some kind of sauce which would make them sticky. Drinks are no problem; you just drink them with straws. Cups or glasses don't work in the absence of gravity, because liquid will come out of a cup when an astronaut accidentally jostles it and float around the cabin. The solution to that problem, which also exists when kids need a drink on a long automobile trip, is available in food stores everywhere: juice boxes.

The written remarks of astronauts about the food indicate, to me at least, that the food is at least reasonably satisfying, though not perhaps as tasty as some earthbound gourmets would like. Astronauts can't resist playing with their food in the weightless environment. Sally Ride writes, "We often share bags of peanuts because it gives us an excuse to play catch, floating peanuts back and forth into each others' mouths. We race to capture spinning carrots and bananas and practice catching spoonfuls of food in our mouths while they twirl in mid-air."[8]

The Bathroom

For some reason, all of my friends, and particularly my son Tom, seem to want to know about a question which William Pogue immortalized as the title of his delightful book, *How Do You Go to the Bathroom in Space?* The answer is: It depends. The all-male Apollo astronauts urinated into catheters and defecated into plastic bags with adhesive rims which they stuck on their bottoms by hand. This scheme was demeaning, messy, and obviously inapplicable to female astronauts. Something better had to come along, though what we have on board the shuttle is not really good enough. The bathroom was the first item astronaut Lodewijk Van Den Berg, who flew in space to grow crystals, mentioned to me when he was asked about the problems of space flight.

The space shuttle's toilet looks, at first glance, like an ordinary, gravity-driven, terrestrial toilet, with a few modifications. NASA doesn't even call it a toilet, preferring the more delicate term "waste collection system" (of course there's an acronym—WCS). There are foot restraints and a seat belt to prevent the user from floating away when it's being used. Suction at the bottom of the bowl pulls the urine and fecal matter down toward the bottom of the toilet. When you're done, you close the lid and flush, just like at home. There's even a place to wash your hands right next to the toilet.

What happens next is not, however, just like what happens at home. As in the old saying, the shit hits the fan. (James and Alcestis Oberg, in their book *Pioneering Space*, pointed out the aptness of this well-known phrase.[9]) When you pull up the handle, called the "gate valve control," a high-speed set of razor-sharp knife blades starts spinning, reaching its operational speed of 1500 revolutions per minute. The solids are shredded into tiny little pieces. If the toilet (excuse me, the WCS) works the way it's supposed to, the solids are slung against the side of the toilet bowl and vacuum-dried to the walls. Some poor soul has to chip the stuff away at the end of the mission. YUCK!

In practice, this device has not worked very well. Parts of it were clogged by the solid waste and had to be cleaned often. It stunk. What's more, in space you don't burp naturally since, apparently, the human body has come to rely on gravity to push bubbles of stomach gas up the esophagus into the mouth. Without gravity, stomach gas either stays in the stomach or comes out the other end, making more smells. And obviously, you can't open the window to air things out. NASA has gone back to the drawing board.

If you're in a space suit on an extravehicular activity (EVA), you have another problem. (Incidentally, I won't use the term "space walk," since it makes the activity sound much too easy.) While it sounds easy to urge astronauts to go to the bathroom first even if you don't think you need to, EVAs often last several hours, and these will become more common as bigger and bigger space stations are constructed. Again, men and women use different systems.

Men go back to the old Apollo-era system of using catheters for collecting urine; the catheter is connected to a plastic bag, which is inside the space suit. They also wear some tight-fitting pants to defecate into if necessary. (Before my kids were potty trained, we called these diapers.) Women wear an underwearlike garment called a disposable absorbent containment trunk (yes, a DACT).

* * *

Clearly, living in space is different from living on earth. Seen in terms of what people can do, though, the differences are less than one might expect. All of the problems discussed in this chapter are basically minor annoyances, curious things that astronauts have to cope with rather than life-threatening difficulties which could pose major obstacles to extended missions. There will be no shortage of astronauts willing to travel to Mars, even if the toilet still stinks. The differences of life in zero gravity have, in fact, allowed NASA to mount a modest but significant research effort in human biology, hoping to use space as a laboratory where you can turn gravity off, in order to better understand how the human body works.

CHAPTER 6

The Dangers of Spaceflight

After all of the concern at the dawn of the space age about the hazards of spaceflight, most of the hazards of the space environment turn out to fall in the category of minor inconveniences. It does seem possible for people to get used to the weightless environment and still work more or less normally. While there are minor differences in the way that the human body works in space, most of these are not life-threatening and are being studied for their own sake, not because they are potentially dangerous.

Yet three problems remain which could derail all hopes for space settlement and might even affect an expedition to Mars: bone thinning, radiation, and the psychological effects of confinement. There are ways around all of these, if the idea is to spend lots of money on a one-shot, symbolic trip to Mars, but they could threaten the viability of a long-term project to settle the inner solar system, particularly if the economics of future space activity dictate that settlement should be at a low cost.

BONE THINNING

Bones are living tissue, not just dead skeletal systems like the steel beams which support high-rise buildings. While an adult's overall bone mass changes only slowly with time, the body uses the bones as a reserve of the critical mineral calcium. Calcium is extracted from the bones when bone cells are absorbed into the rest of the body, and restored to the bones when new bone cells are being laid down. This process occurs most rapidly in growing children, and so children with broken bones can recover considerably faster than adults can. Nevertheless, in a healthy adult, there is a continuing balance between bone formation and bone loss. About 1% of the total calcium reserve is exchanged between bone and extracellular fluid each day in an average adult. As a result, when an adult breaks a bone, it takes a few months for the bone to be regrown.[1]

Changes in bone mass can occur when the balance between laying down new bone cells and cannibalizing the old ones as a source of calcium is disturbed. In older persons, particularly women, calcium isn't absorbed from food quite as easily, and hormonal changes can also upset the balance required for a healthy skeleton. Athletes and others who place increasing loads on their muscles and bones can add more bone mass as a response. Sick people and others who stay in bed for months at a time suffer a loss of bone mass, since their bones and muscles don't have to work so hard against gravity.

Based on all this experience, NASA doctors and their counterparts in the Soviet space program expected that the zero-gravity environment of spaceflight would lead to increased bone loss. What was unexpected was the magnitude of the bone loss and the high volume of calcium which was excreted in the astronaut's urine and feces. Bone loss is quite variable from astronaut to astronaut, but losses of 10% during the three-month Skylab flight seems reasonably typical.

Most worrisome is that bone thinning seems to go on as space flights get longer. During Soviet flights of six to seven months, bone losses are around 15%.[2] While comparing Soviet and American data, and data from different astronauts, has to be done carefully, there is no definite indication yet that bone loss will stop after a certain period of time. Variations from astronaut to astronaut seem to be greater than any overall trend. When the astronauts return to earth-normal gravity, the bones do thicken again, but very, very slowly. There is still concern that the calcium balance would return to zero, where the same amount of new bone is laid down as old bone is resorbed, before the bone loss from space flight had been made up. If this happened, an astronaut would suffer irreversible damage to the skeleton.[3]

The obvious answer to this problem—having the astronauts take calcium pills—won't work. Skylab astronauts ate plenty of milk and cheese, taking in 1210 milligrams of calcium per day, well in excess of the recommended minimum of 800 milligrams per day. The problem is that bone cells are destroyed too fast, and the urine analyses show that. Skylab astronauts excreted calcium faster than they took it in, at a rate of 300 mg per day. This calcium loss was even greater than can be accounted for by bone thinning alone, suggesting that the total body pool of calcium decreased over the three-month flight.[4]

MUSCLES AND EXERCISE

Muscles may be another problem, too. They become weaker in the zero-gravity environment of space since they don't have to do so much work. In space, weak muscles don't hurt you quite so much because you don't have to lift heavy things in order to live there. As a result, space physiologists seem to be less concerned about weak muscles than about bone thinning. In addition, we know that weak muscles can be strengthened and built up again by exercise, and so there is no possibility of irreversible loss of strength in quite the same way that there is for bones.

However, considered in the context of a Mars mission, muscle weakness would create difficulties. If the crew of the first spacecraft that went to Mars were as weak as newborn kittens and couldn't function even in the weaker gravity of Mars, they could do little more on Mars than plant national flags, leave emblems, and then collapse exhausted in their bunks to wait for the return trip to earth. Such a fate could be a disaster if Mars were to be used as a source of important consumable materials (such as water or fuel) for the return trip.

One possible solution to the bone-thinning and muscle-atrophy problems—exercise—might seem obvious, but it isn't as easy as it seems at first sight. Exercise programs in space have had a mixed history. The tiny capsules used in the 1960s had no room for exercise, but the Soviet space station and the American Skylab mission of the 1970s utilized much more roomy spacecraft where a variety of exercise possibilities could be explored.

The boring staple of exercise gyms, the stationary bicycle, has been used on space flights ever since the early 1970s. It's never been popular with the astronauts, possibly because the natural by-product of exercise, sweat, is much more of a problem in the zero-gravity environment of space. Sweat doesn't drip off of the body, and so it simply accumulates on an exerciser's skin. A giant puddle of sweat forms in the small of the back and sloshes around while the astronaut pedals on.

Sweat might be tolerable if there were some way to get really clean in space, but once again the space bathroom is considerably less effective than the terrestrial equivalent. Showers and bathtubs use gravity to pull water from a faucet to a drain. In space, the shower head or faucet can't be placed above the drain, because there is no up or down. Astronauts squirt water over their bodies and suck it up with a vacuum hose. Someone who took a long, luxurious hot shower would find little droplets of sweaty water floating all over the interior of the spacecraft.

In addition, the stationary bicycle only provides limited help with the bone-thinning problem. The bike just doesn't provide the same muscle loading that gravity does, but other devices have been invented which do. A treadmill has evolved as the second staple piece of equipment in the space gymnasiums of both countries. An ordinary treadmill won't work, but a modified one in which elastic straps or bungee cords pull the astronaut toward the treadmill puts a stress on the body which is a little like the normal stress of gravity.

Soviet cosmonauts have tried out a couple of other gadgets which seem to help with the bone-thinning problem and which have the advantage of not interfering with the astronaut's normal workday. A special suit, called the "penguin" suit, contains elastic bands running from the shoulders to the feet and putting stress on your leg and trunk muscles. A different kind of special suit (called the "Chibis" suit) puts lower air pressure on your legs and causes body fluids to sink downward to the legs, in the way that these fluids tend to in earth-normal gravity and don't in space. Similar in function is a suit worn during re-entry and postflight which increases leg pressure, counteracting the force of gravity (this one is called the "G-suit"). Soviet astronauts have also relied on drugs to counteract particular problems in the zero-gravity environment; the American space program has relied somewhat less on drugs.

The fragmentary reports on Soviet astronaut Yuri Romanenko's record-breaking 326 days (nearly a year) in space in 1987 seem to show that these gadgets and exercise regimes do seem to work. Romanenko followed a strict and, according to his reports, "very monotonous" exercise regimen. He did a three-kilometer jog on a treadmill every day, and did his time on the exercise bicycle and in the special suits. Initial newspaper reports indicated that he was quite fit, having only lost 5% of his bone calcium, and jogged 100 meters on the first day after he landed on earth.[5] These results are quite encouraging, though it's not clear whether anyone can overcome these physiological problems or whether Romanenko is an exceptional astronaut. Romanenko's exercise regime was so strenuous that he couldn't work very much, especially in the late stages of the mission. His experience yields mixed results from the perspective of those who hope that routine, long-duration space flight is possible.

Simply saying "exercise," though, is not the answer for astronauts, just as it isn't the answer for physically unfit, desk-bound American workers. Astronauts, like the rest of us, find it difficult to do something unpleasant simply because it's supposed to be good for you in the long run. The amount of time required for exercise, not to mention the effort of getting clean, is rather considerable. Arnauld Nicogossian and James Parker's widely quoted NASA handbook on space physiology states that 1–2 hours a day on the treadmill seems necessary in order to prevent bone loss. So far at least, American astronauts have always been given too much to do during the average space workday. I think it's entirely reasonable that exercise time would be sacrificed so that you could keep to the "mission timeline" that was laid out and still have a few minutes to look out the window at the earth below.

There are, additionally, some difficulties of the space environment which make the usual earthbound approaches to exercise a bit more difficult. Mary Connors, Albert Harrison, and Faren Akins, three authors who produced an extensive compilation of the psychology of spaceflight, write that exercise programs which helps people's "health in some amorphous manner" are rarely continued.[6] Their perception agrees with my own, namely, that people will continue with an exercise program if it is fun, if it develops some kind of skill, or both. Stationary bicycles and treadmills don't fit either of these criteria. Still an additional problem is that many forms of physical exercise on earth involve some kind of competition, and studies of people in confined environments show that they tend to avoid competing with each other. In other words, two astronauts might be reluctant to play racquetball regularly in space because the competition between them would destroy team spirit.

We've not really been very creative about developing space-based exercises; after all, a stationary bicycle and a treadmill are not exactly original pieces of exercise apparatus. Astronauts have very much enjoyed tumbling and acrobatics in space; the Skylab astronauts suggested that all future space stations include such a facility.[7] Perhaps the way around the bone-thinning problem is simply some more creative ways of thinking of exercise, as might be possible on board a space station.

ARTIFICIAL GRAVITY

The brute-force way around the bone-thinning problem has been around for a long while, and is currently called "artificial gravity."

A cylindrical or ring-shaped spacecraft can be spun on its central axis. Someone living on the rim of this spacecraft will be pulled toward the outer edges by a phenomenon known popularly as "centrifugal force." Some physics teachers are uncomfortable with this term because it refers not to a force, but to inertia—the natural tendency of objects to move in a straight line at a uniform velocity unless something pushes on them. It is not a force like the force of gravity or electromagnetism. For our present purposes, though, it serves as a useful label for a phenomenon which could play a very important role in our future in space, and I'll use the label anyway.

What is "centrifugal force," or inertia, or whatever you want to call it? Think of an astronaut sitting near the rim of a rotating space station, reading a book. Because the space station is spinning, at any instant she is moving in a straight line which, if there were nothing to impede her motion, would take her off into outer space. The rim of the space station beneath the astronaut holds her in place, forcing her to travel in a circular path around the space station's axis. The rim pushes on the soles of her feet and on her bottom, and on any other parts of her body which are in contact with the station's rim.

Now think of yourself, and suppose you are sitting in a chair, reading this book in the same way that the astronaut is. Suppose that the chair and the floor of the room you are sitting in were suddenly to turn into jelly, offering no resistance to your natural motion. The earth's gravity would pull you toward it, through the chair, through the floor, and eventually to the center of the earth. What keeps your bones and muscles in trim is that the rigidity of the earth, the floor, and your chair push up on you, preventing you from falling to the center of the earth. These forces act on your feet, on your bottom, and on any other part of your body which is supporting you.

The two situations are really quite similar, from a physiological point of view. In both cases the rigidity of the structure you are sitting in must resist a force which is pulling on a person. In space, that "force" is the tendency of objects in a spinning structure to go flying off into the great beyond; on the earth, that force is the real force of gravity which wants to pull your body toward the center of the earth. The magnitude of the force on the earth is determined by the mass of the earth. On the space station, the station's size and rate of spin determine what the force is.[8] So, if you build a space station right, you can mimic the force of gravity.

How big a station do you need in order to produce artificial gravity? It would seem reasonable that the centrifugal forces on a human being, roughly 6 feet tall, should not vary too much from head to foot—pick an acceptable variation of 10% as a working number. This number gives you a minimal station radius of 60 feet, with a revolution period of about 9 seconds. Many space station designers, as well as the scriptwriters for Arthur C. Clarke and Stanley Kubrick's film "2001: A Space Odyssey," have come up with similar dimensions; it might well turn out that a realistic station would have to be somewhat bigger, so that the astronauts wouldn't get dizzy.[9] While artificial gravity in large space settlements has been studied in some detail as part of the space colony studies of the middle 1970s, space station designers have presumed that it would be cheaper simply to rotate the crew rather than build an extravagantly large rotating station. Of course, crew rotation wouldn't work for the Mars trip.

It seems hard to visualize a station that could be too much smaller, but one reason that artificial gravity is not too popular these days is that such a station is awfully, awfully big. Big structures in space are very expensive. If a big structure like this is needed in order to go to Mars, you not only have to launch it into orbit but you also have to launch all the rocket fuel needed to get it to Mars, match Mars's orbital velocity, and get it back to earth again. Much of this rocket fuel would have to be lugged to Mars and back, and you then need to use more fuel to push the fuel needed for the return trip to Mars. (There are no gas stations on Mars!) The billions of dollars in freight charges begin to mount up, suggesting that it's worthwhile to find out whether there's a better way of dealing with bone thinning than the brute force way of simply putting more metal and more rocket fuel into space.

RADIATION

Think about what the sun looks like at a peaceful, quiet sunset. The soft, soothing yellow rays of the sun keep people warm and make the plants grow. At the right time of the year, they feel soft and warming on your skin. Nothing could be more benign than the soft radiation of the sun.

And nothing could be more misleading than the appearance of the sun at sunset, when considering its possible influence on an unprotected astronaut in space. The sun does emit radiation beyond the narrow range of visible light, such as ultraviolet radiation that causes sunburn and skin cancer. Ultraviolet radiation is the least of our problems in space, since it's stopped quite easily by glass. The major problem from the sun is a form of radiation which isn't even in the electromagnetic spectrum, namely, charged particles.

The sun, such an innocent, yellow disk, contains a tangled, turbulent structure of invisible magnetic fields above its surface. These fields alter the paths of electrically charged particles which pass near them. If the magnetic field pattern is quite tangled, the charged particles can be accelerated so that they travel very fast. For historical reasons, these speeding charged particles are called "radiation," and they can damage human tissue and, if intense enough, even kill you.

The Nature of Ionizing Radiation

This radiation is the same kind of radiation as that produced by nuclear power plants. These power plants and the debris that they leave behind contain a number of radioactive atomic nuclei. When these nuclei decay to make other types of nuclei, they emit charged particles which can travel very fast. A mild dose of such particles can increase your chances of getting cancer in the long term; a heavy dose of radiation from such particles can kill you in the short term. Because of the intense debate about nuclear power plants, the dangers from this type of radiation (called "ionizing radiation" by experts) are probably better understood than the dangers from just about any other environmental hazard.

How does the sun produce this type of radiation? While nuclear reactions do occur in the sun, they occur deep in the solar interior and none of the fast-moving charged particles escape to the surface. Rather, solar radiation is the result of the complex interaction of tangled magnetic fields on the solar surface. These magnetic fields are often found in areas where the sun's surface is quite disturbed, and where sunspots, or comparatively dark areas on the solar surface, are visible. Sunspots are often found in groups, where the magnetic fields thread one sunspot, loop upward in the solar atmosphere, and then descend through another into the solar interior.

The magnetic fields on the sun are sometimes unstable, and can produce a phenomenon known as a solar flare. We don't know exactly how solar flares are produced, but we do know what their visible and deadly (to an unprotected astronaut) by-products can be. Solar flares differ enormously, but the sequence of events in a "typical" flare is something like the following: The tangled magnetic fields, often associated with a complex sunspot region, become unstable and inject a large dose of high-speed particles into the upper layers of the sun's atmosphere. The first visible sign of a flare is often an increase in the amount of visible light emitted by this part of the solar surface; hence the name of the phenomenon. Since it takes eight minutes for light to travel from the sun to the earth, earth-based astronomers can—if they are watching—find out about a solar flare eight minutes after it occurs. In many solar flares, other types of electromagnetic radiation, such as radio waves, are produced; these, too, reach the earth eight minutes after they are generated in the sun.

From the viewpoint of space flight, the important by-product of the solar flare is the burst of high-speed charged particles produced along with the flare. These high-speed particles take longer to reach the earth than visible light does—in some cases, arriving after a time interval of two to three days. A number of science fiction authors have used this time lag to create dramatic situations, where astronauts know that something is about to hit them and have to take evasive or covering action in time. These particles are the ionizing radiation which is one of the dangers of living in space.

What do the high-speed charged particles do to you? They zip through your body, knocking electrons out of atoms, and creating chemical changes in your cells. The amount of damage created is measured in rads (radiation absorbed dose), if you're talking about damage to any kind of tissue or substance, or rems (roentgen equivalent, man), if you're talking about damage to human tissue. At low doses, less than 25 rads or so per incident, there's no immediate effect, but some subtle biochemical changes in your body produce an increase in the chances that you'll get cancer later on in your life. At higher doses, the radiation produces damage to organs which you need in order to continue functioning. Heavy but nonfatal doses lead to such symptoms as diarrhea and vomiting, often accompanied by hair loss, since hair is the fastest-growing tissue in the body. Enough radiation, in quantitative terms roughly 400 rads, can kill you.

The Space Environment

Solar flares can occasionally produce enough radiation to kill an unprotected human being, even if that human being is located near the earth, 93 million miles from the sun. The reason that we're all still here is that the earth's atmosphere provides us with a fairly substantial protective layer. It's a little hard to think of the atmosphere as something massive; air, after all, is the lightest stuff that most of us come in contact with. But when you add up all of the layers of the atmosphere which are on top of you, there's 15 pounds of air above every square inch of terrestrial surface. High-speed particles traveling in space bump into atmospheric particles and slow down.

Above the earth's atmosphere, the earth's magnetic fields also provide some protection against radiation from solar flares and from other sources deep in space. Since this form of radiation is just high-speed charged particles, the motion of these charged particles is deflected by magnetic fields and hence astronauts in orbit near the earth are protected, to some extent, from radiation. This protection does not help astronauts traveling to more distant locations in the solar system, like the Apollo astronauts traveling to the moon or future astronauts on a trip to Mars.

What can be done about solar radiation? All you need to do is to put some mass between human beings and the source of radiation, which for the most severe doses is likely to be the sun. The amount of mass in the earth's atmosphere, 15 pounds per square inch or about 1 kg per square centimeter, corresponds to a layer of rock about 2 meters thick.

While such a shield doesn't sound like much, it is a lot of stuff to haul up into space, particularly if you have to haul it up at great cost from the surface of the earth. Suppose that a space station module or a Mars vehicle were to be about the size of Skylab—our extravagantly large space station which, including all attached components, was 6 m in diameter and 36 m long. Protecting this would take about 7500 tons of rock[10] (about 10 tons per square meter of exposed surface area). It would take about 250 shuttle flights to launch this shielding into orbit if a mission planner were stupid enough to obtain the shielding this way. At the current flight rate, it would take about 20 years simply to launch the shielding for such a mission.

The numbers cited above are somewhat extravagant, because they are based on an attempt to shield the entire spacecraft to the same degree of safety that we have down here on earth. The biggest solar flare in recent history occurred in August 1972; fortunately, there were no astronauts in space at the time. A detailed study based on this flare event suggests that 0.2 m of aluminum shielding would be enough to do the trick, to keep the radiation dose to astronauts below the established limits for people in high-risk occupations.

Another way to reduce the amount of shielding needed is based on a century-old concept perhaps familiar to readers who live in the American Midwest: the storm cellar. Tornados are a significant threat in this part of America, especially in summer. It is impossible to build a reasonable house which can survive if it's in the path of a tornado, unless you happen to just love to live underground. However, it is quite possible to have a reasonably well-protected, small cellar which you can retreat to until the danger is past. A crew of half a dozen people would go bananas if they had to remain cooped up in a room the size of an elevator cage for nine months. However, they probably could stand being in that elevator cage for a day or so.

The scheme, then, is that a spacecraft making the Mars run would contain a small module, probably about the size of an elevator cage, which would be quite heavily shielded from radiation. The rest of the spacecraft would be shielded much more modestly. A rough calculation[11] shows that a 2.5-m-diameter sphere, which would seem sufficient for six people, could be shielded with about 40 tons of material. If you lifted this stuff off of the earth's surface, it would take two space shuttle flights to put the shielding into orbit. Of course, there are cheaper and safer ways of hauling rocks into orbit than using the space shuttle, and you don't necessarily have to procure this material from the surface of the earth. These ideas will be explored in later chapters.

We didn't worry about this radiation shielding in the Apollo missions, and no significant damage occurred. Many Apollo missions flew after the peak of the solar cycle, when intense solar storms are rare. Mission planners did all that they could to plan spacecraft trajectories which avoided regions of space near the earth which have higher concentrations of radiation. As a result, astronaut's career exposures to radiation in space range from less than 1 rem to about 8 rem, well below the lethal dose of 400 rem and roughly comparable to the radiation dose which these people got from diagnostic X rays and from simply being alive: about 2 rem. Persons living in Kerala, India, receive a high dose of radiation from the thorium-rich sands in the region; someone who lived in Kerala for 10 years would get a dose of 50 rem, six times the maximum dose for the astronauts.[12]

Yet we did take chances, calculated risks, in the Apollo mission. The big storm of August 1972 did occur at solar minimum, and fortunately there were no Apollo missions that flew at that time. I've not seen any detailed calculations of what would have happened to the Apollo astronauts had they been in space at the time of the August 1972 flare. A reasonable guess is that the spacecraft probably had enough shielding to protect against a lethal dose of radiation, but that a dose of tens of rems or even more would be likely.

METEORS AND SPACE JUNK

These, long a popular topic in science fiction stories, are really no problem for all but the most visionary of long-lived space colonies. High drama can be created by writing a story in which a meteor smashes into a space vehicle and a heroic crew member has to take immediate action to save the mission, often risking his or her life. However, such scenarios are quite unlikely.

The earth's atmosphere does shield us against the very small meteors, ones about the size of a pinhead, but it is no shield at all against the bigger ones which are found in the older science fiction stories. This size difference creates a confusing terminology (which you don't have to remember; I'm only going through it in order to clear it up). A small, millimeter-sized object traveling through space is called a "meteoroid" when it is just zipping around through space, traveling on its merry way, and not hitting anything. When such a tiny object bumps into the earth's upper atmosphere, it burns up completely, leaving a short-lived vapor trail which is visible from the ground as a shooting star or meteor. Baseball-sized objects are only partially eroded by the atmosphere, and these do fall to the ground (at which time they are called "meteorites"). To my knowledge no one has ever been killed by a meteorite, and there are only a few well-documented cases of one of these ever hitting a house, and so it's reasonable to conclude the chances of any particular house, one of the hundred million houses on earth, being hit by a meteorite are very, very slim. Since the earth's atmosphere has no effect, the chances of a space station being hit by a similar baseball-sized object are also very, very slim, less than one in a billion for a two-year trip.

The small meteorites and pieces of space junk are the ones which hit more often, but don't create much damage. A tiny paint flake from some other spacecraft hit a window on the space shuttle *Challenger* and gouged out a quarter-inch-deep hole, which did not create a leak.[13] Increasing human activity in space has led to a dramatic increase in the number of orbiting objects. Some of these objects are large satellites which travel along known paths and can be avoided, but a larger and growing number are spacecraft debris from spent and exploded upper stages of expendable rockets.

Some attention to this problem in spacecraft design can help cope with this hazard. NASA's usual approach to safety has been redundancy, to have at least two of everything. Designing a space station or colony so that there are a number of independent pressurized compartments would minimize the damage in case a hit did occur. The holes which are most likely to occur are also those which are the smallest, and the easiest to patch temporarily. Meteors should not present a fatal problem to space settlements, or to a trip to Mars.

SMALL GROUPS IN CONFINED QUARTERS

Perhaps the least understood hazard to a trip to Mars, and especially to space settlement, is the psychological reaction of groups of human beings to the rather difficult situation of being locked up in a small can for long periods of time with no possibility of escape. The Mars trip serves as one convenient focus. It's nine months to Mars and nine months back, with the interval of time in between varying depending on mission profiles. There's no way for someone who gets violently upset at his or her crewmates to take a walk outside and calm down, or to go back home for a weekend to gain some perspective on the situation.

For much of the early part of the space program, the psychological aspects of spaceflight were of little concern to the astronauts themselves. During the 1960s, mission durations were two weeks at most, and somebody with the "right stuff" (to use Tom Wolfe's delightful term yet again) can stand just about anything for two weeks. The astronauts saw psychologists as people who definitely did not have the right stuff but who still could bump an astronaut from a coveted flight opportunity because he gave a weird response to some unintelligible test.

Since the 1960s, though, a number of psychologists have been busy studying a variety of groups who end up in situations rather like a trip to Mars or an extended stay in a space station or settlement. Submarine crews, people at Arctic and Antarctic bases, and divers who spent two-week periods submerged beneath 200 feet of water all provided some insight onto the problems. Social scientists Mary Connors, Albert Harrison, and Faren Akins have summarized the results of all this research in a compendium published by NASA.[14]

Even though the missions of the 1960s crammed two or three people in a tiny spacecraft no bigger than the front seat of a very small car, the Apollo astronauts withstood the situation cheerfully. There are many reasons which could explain why the early astronaut group felt relatively little stress. Flights were short, generally a week or less, and people can stand just about anything for a week. These astronauts were almost exclusively military test pilots, who shared a common subculture which placed a high value on someone who followed orders without complaining.

Perhaps the most important reason, however, is that the astronauts of the 1960s were the first group of astronauts, an elite group of men engaged on a heroic, important enterprise. Psychologist Robert Helmreich of the University of Texas, an expert on small groups in stressful environments, has discovered that reactions to confinement can be usefully understood by recognizing the costs and rewards of a particular activity.[15] Someone who is the first to do something like going to the moon feels a tremendous sense of accomplishment, and this person can put up with a lot of grief in order to do what must be done. The 25th, 500th, or 1001st person to go to the moon will feel no such satisfaction, and may well insist on a more satisfying environment with an amount of space a little more comparable to what human beings are used to.

Being the first somewhere may explain one of the most remarkable stories of a small group of people surviving a stressful situation. One of the most dramatic of all the Antarctic adventure stories comes from the Nordenskjold expedition, where three inadequately equipped men spent a six-month Antarctic night of 1903 in a small hut, less than 7 feet by 7 feet, on the tip of the Antarctic peninsula. Around this time, a small group of expeditions had begun to penetrate the pack ice off the Antarctic coast and spend some time in the Antarctic in winter. Yet Swedish explorers J. G. Andersson, S. Duse, and T. Grunden had no intent of wintering over on the continent; they were part of a larger expedition led by Otto Nordenskjold and were simply trying to make contact with another five-man party who had spent the previous winter on an island off of the coast.

Bad weather trapped them in appallingly cramped conditions. They built a tiny stone hut to live in, 7 feet on a side. They killed 100 penguins and many (but not enough) seals to feed themselves and provide heat, and settled down to wait. In the winter, the hut was 10 to 20 below zero Celsius. In the spring, icy water made puddles several inches deep on the floor. For three months, only the flickering of burning seal blubber provided any light at all. Yet they did not go mad.[16]

Psychological considerations, then, may not make so much difference on the first trip to Mars, since the astronauts on the Mars run will benefit from a tremendous psychic reward of being the first to go. Yet in this as in many other respects the trip to Mars is quite different from sending astronauts back to the moon or to a space station in low earth orbit. If we go back to the moon, the next person to step on the lunar surface will be number 13—a superstitious number, perhaps, but not a heroic one. Going to low earth orbit is certainly not a "first," with the possible exception of the first people in particular categories like Christa McAuliffe, the first teacher in space.

An additional difficulty with the Mars trip is that problems are hard to recover from. When Apollo 13's oxygen tanks blew out, they were almost as far away from the earth as human beings had ever been, but it was still possible to bring them back to earth within the week. Such an emergency sprint for home could not happen on a Mars trip. Problems like a crew member becoming uncontrollably crazy must be handled on board.

The need to confine the Mars crew for long periods of time, combined with the difficulty of rescue, suggests that more experimentation with long-duration space stays would be very useful. NASA had some experience with three-month flights on Skylab 15 years ago. The Soviet Union has had considerably more experience, sending pairs of astronauts to their Soyuz and Mir space stations for months at a time. The bits and pieces of stories which emerge from the Soviet experience give some vague indications of difficulties on the very longest flights, but few firm facts on just what the troubles are and what should be done about them have emerged.

Two areas of concern have been identified so far by psychologists studying comparable situations like Antarctic bases: spacecraft design and crew selection. Spacecraft design has been done with relatively little attention to habitability, and inconveniently placed ventilation systems created some minor frustrations in Skylab. Most cabin debris, including important bits of paper like checklists, tends to collect at the exhaust vent, since there is no gravity to cause it to fall to the floor. On Skylab, the exhaust vents were hard to get at, and some important items were nearly permanently lost in the spacecraft's "lost and found" area.

Particularly high on astronauts' lists of desirable spacecraft features are the provision of some kind of private space for each person living in the spacecraft. But how much is enough? How is this private space best shaped in a weightless environment? Soviet space experts James and Alcestis Oberg report that a privacy curtain was originally suggested on the Soviet Soyuz space stations, but that it was never used.[17] Was this really a mistake, or would a simple word in training to the astronauts that it was OK to use the privacy curtain have made life easier?

Crew selection for the current set of space missions generally proceeds in two stages, of selecting the people and then of grouping them together in compatible teams. NASA has paid considerable attention to the crew selection, though they have had the luxury of selecting a few astronauts out of a pool of thousands of candidates. Psychologist Kirmach Natani has identified a new psychological characteristic of "adaptive competence" which seems to be the key factor in selecting astronauts,[18] though intuition may have played a larger role in selecting the American astronauts who have flown so far.

Team selection on both sides has been quite haphazard, with particular astronauts being shifted from one mission to the next with no attention being paid to the consequences for either crew. On the American space shuttle missions, the strong work orientation has given the crew very busy schedules, and there seems to be no difficulty in getting this group of seven people to work together for a week.

Looking ahead, though, selection of individual crew members, and grouping them into teams, might well become a bit more difficult and perhaps controversial. As I see it, the first commercial, self-supporting base in space is likely to be a space station in low earth orbit which will rent space in a zero-gravity laboratory to industrial customers. Some company studying advanced metallic alloys will want to send its own engineer, or team of engineers, on board the space station in the same way that scientists doing polar research send team members to spend the winter at an Antarctic base. Antarctic polar scientists must pass the Navy's psychological screening, and the space station operator (NASA, its Soviet counterpart IKI, or whoever) will probably apply an analogous procedure.

In a small-scale way, this scenario has occurred already. McDonnell Douglas corporation and NASA agreed in the late 1970s to test an apparatus called EOS (Electrophoresis Operations in Space). This apparatus was designed to separate valuable drugs from the chemical stew which contained them. Either McDonnell Douglas could train an astronaut to operate EOS, or NASA could train McDonnell's scientist Charles Walker in how to be an astronaut. The second choice seemed easier, and Walker was the first of many "payload specialists" who have flown on space shuttle missions.

But it is one thing to send a person to space for a week and another thing to put him or her in a space station for three months or even longer. If space is to be settled in any real sense, people like Charles Walker will make up a growing fraction of the group of spacefarers. Space stations will need people whose primary expertise is in metallurgy, crystal growing, astronomy, or something else which can be done in a space station. The crew will become more heterogeneous, with crew members of different nationalities, ages, educational backgrounds, and sexes.

Comparable experiences in the Antarctic suggest that the distinctions between various backgrounds become blurred during a long stay in a remote location. Every year, around 20 people spend nine months in isolated quarters at the Amundsen-Scott South Pole Station. Half of these people are scientists, picked individually by people with experimental equipment located at the pole. The rest are support personnel hired by Navy contractors. Don Kent, a consultant to the Bartol Research Institute, has hired many scientists to spend the winter at the pole. He tells me that the distinction between scientists and others blurs rapidly during the winter, with both groups of people sharing menial chores like shoveling snow.[19]

People of different sexes sharing space on a space station brings up one of the most ticklish issues regarding the psychological dimension of spaceflight: sex in space. Incidentally, women astronauts have demonstrated again and again that they can perform in space just as well as the men can. Astronaut Sally Ride was one of the best at controlling the remotely controlled arm, which knocked a block of ice off of the water vent on her first mission. Kathryn Sullivan has been on a spacewalk or EVA. There is no longer any excuse—if there was any in the first place—to arbitrarily exclude women from the crew.

What's to be done about sex in space? From the point of view of mission planners, Helmreich and his co-authors recommend that NASA treat this problem, phenomenon, opportunity, or whatever you want to call it with "benign neglect."[20] Sending married couples into space won't forestall problems; people break up and get divorced on the earth, even without the constraints of what Helmreich calls "high social density." A similar approach has, apparently, been taken in the Antarctic, where women have been sent to winter over and no one has felt comfortable asking just what went on. After all, why should we? People in space deserve some privacy. . . .

* * *

Living in space is somewhat like living in other isolated, confined areas like Antarctic bases and submarines but it has its own set of peculiar problems. Some of the problems are purely physical, like bone thinning and radiation. When muscles are not stressed, the skeleton is not stressed either. Since the skeleton is living tissue, it becomes modified as a result of the absence of stress and the calcium content of the bones decreases. Space is filled with radiation, high-speed particles that can zip right into your bone marrow and zap the body's red blood cell factories.

There may be a way around these purely physical problems, like designing the right kind of exercise facilities and storm cellars. Indeed, such attention to spacecraft design is part of NASA's Project Pathfinder, an effort to produce the technology needed to settle the inner solar system and send people to Mars. The brute-force way around the bone-thinning problem is to spin the Mars craft and produce artificial gravity; radiation can be coped with by adding mass to the spacecraft. It's not clear whether these inelegant solutions will make the spacecraft far too bulky to fly to Mars, however.

The psychological problems of locking up the Mars crew in a small can and sending them hundreds of millions of miles away from earth are a bit more ticklish. Current experience in space has managed to avoid many of these difficulties; people can stand just about anything for the short one-week missions which are characteristic of the American shuttle program. If we are to settle the inner solar system, where hundreds of people remain in space for considerably longer periods of time, these spacefarers will not have the psychic rewards of being the first to do anything and they may well be a bit more sensitive to the human factors in spacecraft design and crew selection.

All we can do at the moment is to identify the issues which are likely to be important: How much space per astronaut is needed in order to give each person on a space station a sense of privacy? How can this personal space be best designed for a weightless environment? What sort of tests work best for selecting the crew and screening out potentially troublesome people? Much more experimentation with space stations which are commodious and flexible enough to try out different environments is needed before any answers will emerge.

CHAPTER 7

The Necessities of Life

Provided that we can solve the bone-thinning problem somehow, human beings should be able to live in space for extended periods of time, as would be required for a Mars mission and as is certainly required if space settlement or colonization is to happen. However, supporting life in space is considerably more complicated and difficult than supporting life on the earth. The classic needs of life—food, clothing, and shelter—take on a very new meaning when considered in the space environment. "Food," presumably, includes water. It turns out that the most expensive item which needs to be supplied to the space habitats we have now is water, required to keep the human crew and the spacecraft itself functioning. The interior of a spacecraft is conditioned so that the only clothing which is needed is that required to preserve the customary human decency, and so the traditional need for clothing and shelter is transformed into the need to create and maintain a space environment which is suitably cool.

Air has never been on anyone's "classic" (meaning earthside) list of human needs. Its availability on earth is taken for granted. People who have lived in Los Angeles during a smog alert, and who live in other areas with significant air pollution problems, may understand that available air is not necessarily clean air, but even dirty air still has enough oxygen in it and not too much carbon dioxide. All the air in a spacecraft is that which is supplied on board. Humans, like all animals, convert oxygen to carbon dioxide in order to exist. A spacecraft must contain some system, whether chemical or biological, to remove carbon dioxide from the air and supply additional oxygen. On earth, growing plants do this. Space stations, space colonies, then, are in a special place and have special needs.

Putative space settlers face an economic and engineering challenge which is far greater than previous colonizing expeditions. An expedition sufficiently well supplied with seeds and domestic animals could quickly use this modest amount of starting material in order to produce the necessary food in another environment. Clothing, to the extent that it was needed, could be provided from natural materials or from domestic animals. Shelter was built from available materials; in places like Kansas and Oklahoma, where trees were scarce, houses were built from sod.

MAINTAINENCE OF HUMAN LIFE

Table 4 provides some ideas of the types and amounts of material which the average human being requires in order to survive in space for one day.[1] The numbers given are just averages, since different people are different and metabolize food and oxygen differently under different conditions. In addition, while the "need" for wash water can be comparatively minimal, most people tend to use a good deal more than the minimum. The needs for food, oxygen, and carbon dioxide are comparatively straightforward. Food can be in any form which the astronauts will eat, and carbon dioxide and oxygen are simple and familiar substances.

Early conceptions of space food were that everything had to be ground into mush, and that this baby food then had to be packaged into tubes. Mission planners feared that a slice of bread would disintegrate into thousands of little bitty pieces, and that bread parts would find their way into electronic circuitry and affect spacecraft operations. No such dangers materialized. NASA's space shuttle has a pantry which is filled with hundreds of kinds of goodies, the usual sorts of foods that you might find back home. A choice of entrees might include turkey tetrazzini, filet mignon, or barbecued beef. Ice cream was a very popular dessert among the Americans, and borscht was a popular soup for the Soviet astronauts.[2]

People's sense of smell tends to degrade in space, and astronauts have tended to use spices a good bit more than they do on earth. On early Salyut missions, what was supposed to be a six-month supply of garlic, horseradish, and other condiments lasted only a few weeks. Americans also have large appetites for spices, but for different ones like taco sauce and mustard.

In the space shuttle at least, most food is provided freeze-dried. Plastic pouches contain individual servings of particular items, and cooking consists of adding water and heating it. Conventional spoons, forks, and knives do the same job they do on earth.

Water, however, is a critical ingredient, since much of the mass of the foods we eat is water. In fact, most of the mass that's needed to keep a

Table 4. One Person, One Day, in Space

Inputs		
	1.4	pounds of food (dry weight)
	7–9	pounds of water (includes drinking water and the water content of food)
	2	pounds of oxygen
Outputs		
	3.3	pounds of urine
	4.0	pounds of "metabolic water"—exhaled in breathing and evaporated from the skin
	2.2	pounds of carbon dioxide
	0.4	pounds of solid waste

human being alive is in the form of water. The total mass of stuff needed to keep one person alive for one day is a little over ten pounds, and about eight of those pounds are in the form of water.

The question of water supply is a bit complex, since the quality of the water which is required for various needs is different, depending on what one wants. Water used for drinking and in food preparation must, for both esthetic and health reasons, be quite pure. The standards for wash water, while not quite as stringent as the standards for drinking water, are still quite high. Water dissolves just about anything, and small quantities of noxious pollutants can make water unfit for use in space.

Current spacecraft also use water for other purposes, just to keep the spacecraft going. For example, the space suits that astronauts use on extravehicular activities (EVAs) boil water to keep the astronaut cool. On EVAs, astronauts tend to spend a good deal of time in the sun, just because it's easier to do work when sunlight illuminates whatever it is the astronaut is working on. Heat from the sun plus whatever energy the astronaut generates from simply moving around must be dissipated somewhere, and the usual terrestrial mechanism of evaporating sweat into the atmosphere is not available. As a result water vapor boils off the spacesuits in order to keep the astronauts cool. Water is also used in the space shuttle's thermal control system; it circulates throughout the spacecraft, keeping it at a uniform temperature.

In some cases at least, a spacecraft can actually produce water as a by-product of its operations. Electrical power in the space shuttle is produced by fuel cells, in which hydrogen and oxygen are chemically combined to produce water and electricity. This system provides an extravagant amount of water, about 72 kilograms per day. The water is used by the crew and by the spacecraft's cooling system, but there is still some left over and the excess is simply dumped overboard. Thus the usage of water by the space shuttle, which corresponds to 10–15 kg per person per day, is unnecessarily extravagant and should not be considered to be a model for future space stations.

RECYCLING INSTEAD OF CONSUMING

On earth, of course, the necessities of life are not simply consumed and thrown away. Rather, a number of global cycles transport these precious materials through the global environment and reuse them. Water which you flush down your sink eventually reenters the global water supply, whether it is released into the soil as part of a septic system or discharged into a river by a sewage treatment plant. This water evaporates into the atmosphere, a process which purifies it still more, and falls to the earth's surface as rain, becoming once more available for human consumption and use.

At the height of the environmental movement, environmentalist guru R. Buckminster Fuller coined the phrase "spaceship earth" to emphasize the closed nature of the earth's ecosystem. This phrase, along with pictures of the earth taken from space, may have made some people conscious of the fact that you never throw garbage away, you merely put it someplace else and hope that it will stay there and not bother you.

While Fuller may have been able to communicate the essence of ecology by using the phrase "spaceship earth," the phrase couldn't be more inappropriate when it comes to describing spaceships. Today's spacecraft, at least, are totally unlike the earth. They need continual resupply from the earth's surface and dump lots of junk overboard. The space shuttle is surrounded by a little cloud of water vapor, the result of "venting" of urine and surplus water from the fuel cells. Trash piles up too, but the space shuttle is roomy enough and the missions are short enough that the trash pile doesn't get in the way.

The Soviet space stations are in orbit for longer periods of time, but they too are far from earthlike in their recycling capabilities. The Soviet space stations are resupplied every couple of months by a robot Progress spacecraft. New food, water, and other supplies are extracted. The Progress's cargo hold becomes a garbage can for paper, toilet bags, food wrappers, and anything else that the Soviet astronauts need to get rid of. Progress then plunges back into the earth's atmosphere and burns up, trash and all.[3] The Soviet Salyut space stations have been such prodigious generators of trash that the bimonthly garbage collection provided by the Progress spacecraft is not enough. Periodically, little packages of who knows what are dumped overboard and, like the bigger Progress spacecraft, end up in the atmosphere and disintegrate.

We can do better. Or at least we ought to be able to do better. I have often seen students and other interested listeners be very impressed when they are told that astronauts drink water which is produced by the space shuttle's fuel cells. These fuel cells combine hydrogen and oxygen to produce electricity and a rather unusual "ash"—nearly pure water. People who are impressed by the space shuttle's system have the right idea; it is desirable to reuse water in space. We do want to transform spacecraft into earthlike spaceships, not voracious, once-through consumers of water and generators of trash. However, fuel cells are a step in the wrong direction. The space shuttle uses them because it would be time-consuming and complicated to deploy solar panels and refold them one week later. However, space stations can be powered by solar panels. Fuel cells will only provide an incidental source of power; if astronauts count on them for drinking water, they should be considered part of the life-support system, not as part of the power system.

What we seek to do is to reuse as much of the necessities of life as we can. Spacecraft designers use the term "closed-loop" life support to describe the ultimate goal. An ideal spacecraft would resemble the earth in that once it was built and initially supplied, no new material would be needed from the outside and nothing would be dumped overboard. Some time in the future, we can approach that ideal. At the moment, spacecraft designers and others studying this problem are only partially trying to close these various loops, to reuse some major fraction—say more than half—of these precious quantities that astronauts require. The three loops which require closure are the water loop, the air loop (e.g., oxygen and carbon dioxide), and the food loop. While there are important interconnections between all three loops, the technologies are different for each one.[4]

The Water Loop

Apart from any excesses provided by spacecraft systems, the water generated by human activities comes in basically three forms: sweat and breath (which appears as cabin moisture), wash water, and urine. Humans require water for drinking (including food preparation) and for washing. The spacecraft itself may require water to make its cooling systems work.

To be sure, water is water, but the purity required is different depending on what you want to do.

The easiest water to recycle is the cabin moisture, though it is not absolutely pure. Once it is condensed on a dehumidifier, well-known techniques for water filtration can get most of the pollutants out. There is no question that this recycled condensate is suitable for wash water. The Soviet space program uses this condensate for drinking water; American space planners are somewhat more uneasy about that.[5] Because American spacecraft have, so far, supplied an abundance of water from fuel cells, no decisions have had to be made about whether it's wise to drink cabin condensate.

What's a little more challenging, both to technology and to the squeamish psychology of future spacefarers, is what you do with the urine. Urine constitutes a little less than half of the water which an average astronaut gets rid of in each day. A system which recycled everything else perfectly and threw all of the urine out into space would still consume 3.3 pounds of water per person per day, far too much for a long-lived space mission. A life support system which doesn't reuse urine is far from a completely, or even almost completely, closed system.

The problem with urine is that it contains a fairly high concentration of chemicals like urea which are very hard to get rid of by the normal processes of distillation. A number of investigators have tried to build gadgets which transform urine into acceptably pure drinking water. While, from a purely technical viewpoint, the results seem promising, the investigators don't seem to be able to persuade either themselves or anyone else that drinking distilled urine is OK.

Of course, we drink reprocessed urine all the time. The water cycle on the earth is more than just a physical cycle of evaporation and filtration. Soil contains loads of organisms which just love to consume the urea in urine and purify it. Purely physical mechanisms such as evaporation and distillation just can't do as well as biological systems can, suggesting that we need to incorporate plants into the water loop. When we consider some of the other loops which a long-lived space station or colony or base needs to close, involvement of biological systems becomes even more crucial.

Because it involves such a large amount of mass per person, closing the water loop is one of the crucial items in designing any spacecraft which is intended to house human beings for extended periods of time. If the water loop is open, then it will cost something like $80,000 per day or $30 million per year to keep a single astronaut alive, if the water is ferried up from the earth at a cost of $10,000 per pound. Reuse 90% of the water, closing 90% of the water loop, and the cost plummets to $3 million per year, almost within the range of private enterprise.

The Air Loop

The Soviets' reuse of cabin condensate and the American use of fuel cell water are beginnings, at least, in the closure of the water loop. Spacecraft of both countries, however, have used a generally open system for air. Simple consideration of the weights involved do suggest that the water loop is the place to start, because the average astronaut uses several kilograms of water and only one kilogram of oxygen (the oxygen is exhaled in the form of carbon dioxide) each day.

In the American space program, oxygen is brought up to the spacecraft in pressure tanks and allowed to bleed into the cabin air. Getting rid of the carbon dioxide has, in every case except for Skylab, involved a very simple process. Small canisters of lithium hydroxide (LiOH, pronounced "lye-oh") pellets and activated charcoal sit in a compartment on the space shuttle floor. Fans blow cabin air past these canisters. Carbon dioxide and other pollutants just physically stick to the LiOH and charcoal, being adsorbed onto it. Every 12 hours, an astronaut has to replace one of the canisters with a fresh one and throw the old one onto the steadily growing trash pile. The Soviet program uses a somewhat similar system, where potassium superoxide simultaneously provides a source of oxygen and absorbs the carbon dioxide. In neither case are any of the oxygen atoms in carbon dioxide recycled into the atmosphere.

There are purely chemical processes which can strip the oxygen atoms off of a molecule of carbon dioxide and dump them back into the atmosphere. In both cases, hydrogen atoms are required as catalysts. The simpler one, called the Sabatier process, forces hydrogen and carbon to combine to form methane, releasing oxygen and heat. Methane, a colorless, flammable gas which is a major component of natural gas (used for cooking and heating), is of no use whatsoever on a spacecraft; a methane buildup could explode the spacecraft or asphyxiate the astronauts. So, once again, recycling is not perfect and something is dumped overboard—in this case hydrogen. The Bosch process transforms the hydrogen into water and produces carbon trash which is also thrown overboard. It's likely that one of these chemical processes will be used on the American space station.[6]

But nature can do better. A green plant makes some use of the carbon left over when you strip the two oxygen atoms from a carbon dioxide molecule, using the carbon as food. The chemical reactions which make up the photosynthetic pathway in green plants are not completely understood, and furthermore are quite complicated, so we can't just simply assume that a future spacefarer will be able to build some glassware, mix some chemicals, and do what a green plant will do. It's far easier to try to grow a green plant in space, and this is the goal of some very active attempts to create an earthlike spaceship.

However, in the past, it has not made sense for NASA to do better. To close the water loop or the air loop, NASA has to spend money up front designing an appropriate system and then spend more money to launch the filtration or distillation apparatus into orbit. All of this effort and expense might not make sense for a short space shuttle flight, where it's probably cheaper to pay the weight penalty of lugging extra LiOH and oxygen canisters into orbit instead of designing, testing, and retesting some novel process.

The "save money at any cost" approach which has led to completely open systems on the shuttle is, however, rather shortsighted. NASA's bureaucratic heritage has caused it to focus on the next big program and not look beyond it. We are now looking beyond one-week shuttle missions and thinking about flights to Mars, but NASA is faced with no real experience with any of these systems. Even the simple ones which use only physics and chemistry rather than biology to reuse precious air and water exist only as ideas on paper and not as hardware which has been tested on the ground, or flown in space. American engineers turn up their noses at the idea of drinking recycled cabin water, while reportedly the Soviets do it. What basis do we have for this conclusion? Have we ever tried to recycle cabin moisture to see what it tastes like?

We do not yet know how complex the life-support system on the space station will be. The most detailed station design that I know of is the "reference configuration" devised by a group of about 80 NASA engineers in 1985. The document prepared by this group bravely talks about drinking cabin moisture, even allowing for the possibility of drinking distilled urine, and converting carbon dioxide to oxygen without dumping anything over the side. However, no details of exactly how this is to be done are provided; indeed, the document explicitly states that "the baseline processes for CO_2 removal, CO_2 reduction, O_2 generation, and water reclamation have not been selected."[7] In other words, astronomers who hope that their telescopes won't have to look through a little cloud of not-so-pure water vapor surrounding the station simply have to hope that these brave goals can be met.

For once, acronyms illustrate what the challenge is. The small NASA program investigating recycling, which has grown to be a part of the Pathfinder project investigating the technologies needed for a 21st-century space program, seeks to devise an object called a CELSS, which is now called a Controlled Environmental Life Support System. That particular acronym could refer to a chemical system. A second stage, also a CELSS, is a Controlled Ecological Life Support System. The eventual goal of this project, with the same acronym, is a Closed Ecological Life Support System. What these acronyms really mean is controlled bureaucratic ambiguity (CBA)!

The space station reference configuration aims at the first definition of CELSS, where the environment is controlled rather than closed. The planning document explicitly excludes, at least at the start, any attempts to close the loop more completely by closing the food loop and bringing biological systems into the game. They say bluntly that "a completely closed ECLSS [now called a CELSS] is not viable at this time because of the high cost and risk of developing food regeneration technology."[8] Again, if your goal is to orbit a space station as soon as possible and with the lowest possible up-front costs, this attitude makes sense. That approach, though, won't help us get to Mars.

CLOSED SPACECRAFT ECOLOGIES

An open life-support system is clearly wasteful on board a spacecraft. Attempts to provide better living in space through chemistry alone still leave us far from a closed life-support system because some waste is still present. As a result, a small but growing number of scientists, some led by Phil Quattrone of NASA's Ames Research Center, some working for private companies under contract to NASA, and some working on an independently funded Project Biosphere, have for many years been studying various ways in which we can emulate all or part of the closed ecological cycles which take place on planet earth. This project aims to build a real CELSS, a closed system, using the most futuristic definition of this all-purpose acronym.[9]

The earth is a complicated, interrelated ecosystem, and attempts to emulate the earth's enviroment inside a closed, small spacecraft have to go far beyond simply having a tray of algae growing beside the pilot's control panel. In nature, nutrients, water, air, and energy are all freely available to a plant in the sense that the plant's use of these quantities may be balanced by some compensating reuse that takes place far away from the plant, in a very different environment. For instance, rainfall in a tropical forest does not have to be balanced by evaporation in that same place, for the excess water can run off to the ocean and evaporate there. A closed system has to contain a variety of environments so that all of the various processes which occur in different places on the earth can also occur in space, or the particular types of plants and recycling processes have to be selected very carefully.

Another advantage of the terrestrial ecosystem is the earth's large size. Environmental disasters in one place need not be fatal. If you overwater (or underwater) all of your house plants and they all die, you won't die too because of a shortage of oxygen in your home. The oxygen which you breathe is not generated only by your house plants. Sheets of marine plankton thousands of miles away from your home can play a role in maintaining the balance between oxygen and carbon dioxide in your home. In space, however, a completely closed or even almost completely closed system has to be sufficiently stable so that such disasters won't disable the spacecraft.

Prototype Space Gardens

The CELSS project has gone beyond—but only just barely—pencil-and-paper designs. A steel chamber with a volume of 80 cubic meters contains a completely controlled environment at the Kennedy Space Center in Florida. Lamps provide the light source, and a variety of crops were grown in 1985–1986. While all this research is going on on the ground rather than in space, it at least provides some experience on the kinds of plants which survive best in a prototype closed system. What crops to grow? The choice is not obvious. Plants in space will have to grow without gravity, and in general it's important for a plant to be able to distinguish up from down so it can point its leaves toward the sun and its roots toward the ground. The short duration of space shuttle flights makes it difficult for NASA biologists to study experimentally the plant growth cycle in space, and no such experiments were done on the longer Skylab flights of the mid-1970s.

Again, the Soviet Union's space program has marched ahead, perhaps because they have had an eye on a Mars expedition for some time now, but perhaps because they have found that gardening is something that astronauts genuinely like to do. Early experiments in the 1970s were failures; the plants simply wilted and died. In 1979, a tulip in bloom was sent into space, and died the second day it was in orbit. A year later, an orchid was sent up, and its petals fell off. Finally, in the 1980s, a device called Phyton was developed that worked. James and Alcestis Oberg, two American experts on the Soviet space program, attribute its success to a different nutrient medium, to filters which keep the plants isolated from contaminants in the cabin air, and to artificial illumination of the plants. A spacecraft in low earth orbit, as the Soviet space station is, has a bizarre day/night cycle, with sunrise occurring about every 90 minutes. Evidently considerable help from light bulbs is needed to prevent the plants from becoming confused.[10]

However, growing a few plants in orbit is a long way from establishing an independent, self-sustaining food system. Any engineering project must follow a long, tortuous path from an idea in someone's head to operational reality. The hazards of the space environment, the need to exhaustively test anything which is a vital support system for human life, and the NASA bureaucracy make it even harder to shepherd an idea from conception to its birth as a real operating space system.

As a result, while no insuperable technological obstacles seem to stand in the way, the idea of a completely closed ecosystem, where the food loop, water loop, and air loop are all intertwined and closed and where water dumps are a thing of the past, remains in the concept formulation stage. NASA planners have invented the term "concept formulation" to describe an idea which has been fleshed out to the point where you have some idea of what it's going to cost, but where no metal has been bent. Many ideas which pass through the concept formulation process turn out to be unworkable, too expensive, or both. An earthlike spaceship is beyond the science fiction stage, but only by a little bit.

Project Biosphere

An independent, privately financed project currently being conducted in the American Southwest may be the key link between isolated experimental space gardens and a truly closed ecosystem. The goals of this project are not primarily to help the space program, but rather to understand how the global ecosystem works. In Project Biosphere, a small number of volunteers will live for a couple of years in an ecosystem which is completely isolated from the terrestrial ecosystem.

The biosphere is a sealed off piece of Arizona desert which will contain lakes, ponds, fish, insects, plants, and people. Its builders have made every effort to include all of the necessary ingredients inside the plastic dome which isolates the biosphere from the outside world. The global ecosystem which has provided us with closed water, air, and food loops for the last 4.5 billion years includes millions of species of different biological creatures. The principal objective of the biosphere project is to test our understanding of the way that these creatures interact in an ecosystem by creating an artificial one.

In the late 1980s, half a dozen volunteers will enter the biosphere, close the door, and begin living literally in a goldfish bowl for a couple of years. Their interactions with each other and with their environment can serve as a test bed for future space colonies.

OTHER NECESSITIES OF LIFE

I have concentrated mostly on the biological necessities of life because these vital ingredients are common to all space habitats that people talk about. The most critical one is water. Oxygen can be provided from water by dissociating the hydrogen and oxygen if a source of electrical power is available. Food is required in somewhat smaller quantities, and there is reason to believe that at sometime in the 21st or 22nd century spacecraft can provide a significant fraction of their own food. However, in considering what the true cost of life in space is, there are some other vital items which are likely to be needed for the forseeable future.

Shelter

For the astronauts, shelter is provided by the spacecraft itself. Spacecraft in low earth orbit, shielded by the earth's magnetic field from the space radiation environment, are lightweight craft built from light metals. It is very likely that the metals which make up the spacecraft will be obtained from the earth for the forseeable future.

A spacecraft going beyond low earth orbit will find itself in some environments which are not as benign to human life, where some form of radiation shielding will be essential. This shield does not have to be made of some exotic, ultrastrong and ultralight material. A radiation shield works by putting so many kilograms per square meter of stuff between you and the source of the radiation, and it doesn't matter what the stuff is made of. The mass in the radiation shield makes up a very large fraction of the mass of a spacecraft going to Mars, or a space settlement orbiting the red planet.

The eventual possibility of space settlement leads us toward the idea that it might, at some time, make sense to obtain the necessities of life in space from someplace other than the surface of the earth. Lifting material out of the earth's strong gravitational field costs a lot, and with the necessary infrastructure in place it might be easier to get this material from somewhere out in space, if the right stuff is there. Material to make radiation shields is particularly attractive in this respect since any kind of rock will do. Space manufacturers will need to crush the rock, make concrete out of it, and wrap the astronaut's storm cellar in a concrete cocoon. A brief report from a NASA contractor indicated that you can indeed make satisfactory concrete out of lunar soils.[11]

The major ingredient in concrete which is not obviously available on the moon is our old friend, water. However, for reasonable concrete mixtures such as that proposed in connection with a lunar base, water is only 4% of the total. (Most of the mass of concrete is "aggregate" or rock; in this proposed mixture the cement paste is about 12% by weight of the total, and water is a small fraction of that.[12]) Perhaps it makes sense to use lunar material for shielding; another possibility is to process the lunar soil into dark glass bricks.

Another sheltering material needed in space is the inert ingredient in the earth's atmosphere, nitrogen. Pure oxygen atmospheres were tried in spacecraft in the early and mid-1960s. The disastrous Apollo 1 fire put these experiments to a stop. In the pure oxygen atmosphere of the first-generation Apollo spacecraft, a tiny electrical short circuit immediately became a violent conflagration and incinerated three Apollo astronauts. In an atmosphere like the earth's, the smaller fraction of oxygen makes it harder for fires to spread rapidly. The investigation into the Apollo fire made it clear that some kind of inert gas is needed to make the cabin safe. In space, that inert gas is definitely nitrogen.

However, the nitrogen slowly leaks out of the spacecraft, no matter how carefully leaks in the spacecraft have been sealed. Leakage rates are quite low at this point; the space shuttle leaks nitrogen at the rate of about 3.5 kg of nitrogen per day, and projected numbers for the space station are slightly less at 2.3 kg per day.[13] On a per-person basis these numbers are considerably smaller than the needs for water, but over the years they will add up, particularly if we succeed in partially closing the water and food loops. At some time, the future needs of space settlements may well include a nitrogen supply.

Putting It All Together

What's needed to stay alive in space? It depends on how hard we work to try to close the water loop, the the air loop, and the food loop. The way we live now, the answer is an extravagant quantity of almost everything. The way that we could live in the future, we could need far less.

Table 5 illustrates what's needed to keep a crew of six alive in space for a one-year stay. I've given numbers corresponding to three different scenarios, only one based on any detailed spacecraft design. The "space shuttle" scenario is life in space at its most wasteful, based on current space shuttle operations. Short missions and the use of fuel cells dictate that all three loops are wide open and that lots of water is simply used to maintain the spacecraft. In that scenario, the 18,000 kg of water needed for one year of flight operations excludes the water used by the crew.

The "partially closed" scenario is based on the "reference configuration" of the space station, which should represent a reasonable guess at our technological capabilities in the early part of the 21st century. I've assumed here that the goals of the 1985 design are only partially met in that the water loop is 75% closed and the air loop is 50% closed. In other words, 75% of a human being's water output each day can be reclaimed and reused. The food loop is, as is the case now, completely open. The "Space 2100" numbers are optimistic guesses; here I've assumed that the water loop is 95% closed, the air loop is 95% closed, the food loop is 90% closed, and that spacecraft are 100 times tighter than they now are. In order to avoid the illusion of precision, I have rounded all figures, even though we know what the numbers are for the space shuttle with considerably more accuracy.

It is evident that for any reasonable guesses about future technological developments, water is going to be the key ingredient required to sustain life in space, but not the only necessity of life. In any of the three scenarios, and certainly with the space shuttle, water consumption dominates the amount of mass needed to keep human beings alive in space. Oxygen (which can be obtained from water by a simple chemical process, if energy is available), food, and nitrogen are all required in roughly equal amounts, in all cases in lower quantities than that required for water.

If that stuff has to be all lifted up from the surface of the earth, at a current cost of roughly $10,000 per pound, then the cost of maintaining a six-person crew, even in a partially closed ecosystem, is going to be a substantial $100 million. Even with fairly complete closure of all the

Table 5. Necessities for Life in Space[a]

Supplies	Space Shuttle	Partially Closed	Space 2100
Water for:			
Drinking	8,000 lb	2,000 lb	400 lb
Washing	4,500 lb	2,200 lb	450 lb
Spacecraft operations	36,000 lb	—	—
Oxygen	4,000 lb	2,000 lb	200 lb
Food (dry weight)	2,500 lb	2,500 lb	250 lb
Nitrogen	2,000 lb	1,500 lb	150 lb
Total consumables	57,000 lb	10,200 lb	1,450 lb
Total excluding spacecraft operations	21,000 lb	—	—
Cost per year ($10,000 per pound)	—	$100 million	$15 million
Radiation shielding	—	40 tons	40 tons
Water component—if concrete	—	7,000 lb	7,000 lb

[a]Projected needs for a crew of six for one year.

loops, the cost of $15 million adds up to about $2 million per year per person, an extravagant life style.

* * *

These large numbers which emerge from even an optimistic exercise to try to estimate the cost of maintaining life in space suggest that one way to reduce the cost of living in space is to obtain some of the resources from space itself. The short, 100-mile gap which separates the earth's surface from low earth orbit is, in terms of rocket fuel, a huge chasm, especially when compared to the energy costs of getting to other places in space. In the last several years, a number of writers have begun to think about other places to get these scarce resources, and those will be discussed in the next chapter.

PART III

Space Resources

Keeping someone alive in space requires a significant initial investment in a spacecraft, as well as continual resupply in the form of resources. For many years, scientists and science fiction writers who tried to forecast our future activities in space more or less assumed that these resources would be supplied from the earth. A significant change in our thinking about our future in space has occurred in the last decade. Planners have realized that the earth may not be the most logical place to obtain the resources needed to build spacecraft and resupply them. Perhaps these resources could be obtained from bodies in space, so that we don't have to burn up a ton of rocket fuel to lift a mere hundred pounds of water from the earth into orbit. Water and other useful stuff obtained from bodies in the inner solar system are generally called "space resources," even though they are not obtained from empty space itself but from solid bodies in the inner solar system like the moon, Mars, the satellites of Mars, or small asteroids orbiting near the earth.

Largely but not exclusively as a result of 30 years of space exploration, we know considerably more about solar system objects than we did when the space age began. At the present time, there is a vigorous debate within the space community about where we should go once we build a space station in low earth orbit. Is the next target the moon, Mars, or the asteroids? You might first think of the moon, since it's closest, but distances in space are deceptive. The basic mechanics of rocketry suggest that, in some senses, the moons of Mars are closer to low earth orbit than the earth's own moon. However, trips to the moon can occur much more frequently.

Where do we go next? Mars has a great symbolic value as a goal for humanity; were we to go to the moon, it would be going back to where we've been already. If space settlement is to be in our future, though, a more logical way to approach the question of our next destination is to ask what destination can provide the resources we need to sustain life in space at the lowest cost. We need to ask just what sorts of materials are available from the natural destinations in the inner solar system, what use we might make of them, and how much it costs to get there. Such consideration yields some rather surprising answers, in that the most promising destinations are not large, familiar objects like the moon and Mars, but rather are small bodies like the Martian satellites Phobos and Deimos, and various small bodies orbiting in the inner solar system, the near-earth asteroids.

CHAPTER 8

Getting around the Inner Solar System

Why should we look to space itself for the resources that we need? While only a few hundred miles separate the earth's surface from a spacecraft in low earth orbit, the earth's gravity makes these few hundred miles some of the most difficult territory to cross in the solar system. People are used to using distance alone as a measure of how difficult it is to travel from one place to another, but this way of thinking, very useful in considering most places on earth, fails utterly to describe the difficulties of space travel, especially space travel by robot spacecraft without people in them. It takes less rocket fuel to send an automated spacecraft to the surface of Mars than to the surface of the moon. In this sense, Mars is "closer" to the earth than the moon is, though in terms of distance Mars is well over ten times as far away even when it makes its closest approach to the earth.

Another consideration as we look to the 21st century is that there are two propulsion technologies which could make it much easier—and cheaper—to get around the inner solar system. Because space is empty, you can accelerate a spacecraft either with a lot of gentle nudges or with a short-lived, powerful rocket blast. The gentle nudges won't get you off the surface of the earth, but they can bump you from one orbit to another. These technologies are not mature by any means. Solar sailing has barely gotten from people's imaginations into serious calculations, not to mention tests of hardware. A different type of propulsion scheme which uses electricity to accelerate the rocket exhaust has been tested on a small scale. If these technologies live up to their promises, they could become the Conestoga wagons of the 21st century.

THE PECULIAR GEOGRAPHY OF SPACE

Space is different from the surface of the earth, in many ways. The vacuum, the sunlight which fries a spacecraft's sunny side and freezes its dark side, and the peculiarities of the weightless environment have been understood for decades. A less familiar, but no less strange characteristic of space is related to the absence of air in space. Once you push a spacecraft in a particular direction, it simply keeps on going and going and going, orbiting forever within the solar system. With most of the mass of a spacecraft being fuel and oxygen, the cost of getting somewhere is more closely related to how hard a spacecraft has to be pushed to reach its destination, not how long the trip takes or how many miles separate the trip's origin and destination. The concept of distance takes on a new and very different cast in the context of outer space. These are new ways of thinking, and I'm still not sure that many in the space community have fully digested them or understood their implications.

Geography on Earth

Geography, a discipline which flourished in the 19th century and faces a bit of an identity crisis now, takes on a rather peculiar cast when you try to apply it to outer space. The elementary school version of geography, the aspect which most of us are familiar with, was described by one of my daughters' teachers as being simply map-making and map-reading. My neighbor, a geographer, chuckled at the idea; maps do play an essential role in the field, but geography is more than map-making. Behind the mental discipline of map-reading is something more fundamental—developing a sense of place and distance. The sense of place is what the field of geography is all about.

On the earth, what is meant when someone says he or she lives in a particular place? Most of us identify the place that we live as a neighborhood with a name. It's located in various larger units: perhaps a large city or a county, then a state, then a country. You can describe the type of place that you live in by characterizing it as being urban, suburban, or rural. Its physical characteristics can be classified as being hilly or flat, rainy or dry, hot or cold, and so on. One of the pleasures of travel results from the way that people do live in so many different places and adapt to them in fascinatingly different ways.

While places on the earth have many different characteristics, the concept of distance is, in general, really quite simple. Two-dimensional maps can, almost everywhere, do a very nice job of illustrating how hard it is to get from one place to another. The geometrical distance between the two places on the map is often a reflection of how hard it is to get from one place to another. There are exceptions. In America's frontier past, natural barriers like mountain ranges or natural transportation routes like major river systems alter one's perception of a two-dimensional map, but in ways which are relatively easy to understand. The contemporary equivalents of these natural barriers are traffic lights and overloaded major highway intersections. Still, these natural barriers are easily understood mentally, as being simple additions to a two-dimensional map.

Space Places

In terms of human habitability, outer space offers a stark contrast of sameness, set against the large variety of terrestrial neighborhoods and communities. Virtually every place you would want to live in, work in, or send machines to in space is a very hostile environment. The few planets and satellites which do have atmospheres have unbreathable ones. In the rest of outer space, you have no air, no water, a sun on one side which cooks your spacecraft and darkness on the other side of your spacecraft to freeze it. Radiation is a hazard everywhere.

Space may seem limitless, but there are several special, preferred locations in space where a space colony or space station might usefully be put. Most of our space activity so far has been carried out in low earth orbit, a place where spacecraft orbit a few hundred to a thousand miles above the earth's surface. (Of course, there's an acronym: low earth orbit is LEO.) A spacecraft in low earth orbit circles the earth every 90 minutes.

Putting a spacecraft in low earth orbit requires a considerable amount of effort. An orbiting spacecraft travels at a speed of 8.6 km/sec. When you express this velocity as 18,000 miles per hour it becomes a little bit clearer just how fast spacecraft are moving. The speeds of high-performance aircraft are often expressed as "Mach numbers," where an aircraft traveling at Mach 2 is traveling at twice the speed of sound. Since there is no air in space, sound doesn't travel in space, and it's not strictly correct to refer to Mach numbers. However, a spacecraft on its way to low earth orbit ends up traveling at Mach 25 when it zips through the earth's atmosphere.

The orbital speed of 8.6 km/sec gives the craft a sufficient velocity sideways so that it falls around the earth rather than directly toward the earth. The only way we know how to give a spacecraft such a high velocity is to put it on top of a big rocket and use lots of rocket fuel to get it into orbit. The space shuttle, at launch, weighs about four million pounds; the orbiter, with its maximum payload of about 65,000 pounds, weighs about a quarter of a million pounds, less than 10% of the weight of the entire stack. The NASA term "stack" refers to the entire assembly of machinery on the launch pad: fully loaded fuel tanks, booster rockets, orbiter, and cargo. The payload is the cargo that you care about—the equipment or people that are the reason for going into space in the first place. The exorbitant fuel requirement needed to get to low earth orbit makes spaceflight very expensive.

There's a special orbit around the earth called the geosynchronous orbit (GEO) which is of particular interest to the communications business. A satellite in geosynchronous orbit, 22,300 miles from the earth's center, takes exactly 24 hours to orbit the earth. If this satellite orbits over the earth's equator, it will remain fixed from any vantage point on the earth's surface. Someone on the earth can receive signals from this satellite by pointing an antenna in one direction and leaving it there; there's no need to track the spacecraft and follow its orbit. So many communications satellites are in geosynchronous orbit that an international organization has been formed to assign parking places in this orbit to various countries and organizations.

Getting from low earth orbit to geosynchronous orbit requires an additional velocity boost. The total amount of change from low earth orbit is a bit complex, because once you reach geosynchronous orbit you need to do an additional rocket burn to put the satellite into an exactly circular orbit. When you add it all up, a velocity change of about 4 km/sec is required to go from low orbit to geosynchronous orbit.

In terms of the amount of effort involved, then, it takes twice the velocity change to go from the earth's surface to low earth orbit, a distance of a few hundred miles, as it does to cover the 20,000 miles which lie between low orbit and geosynchronous orbit. The relative distances between the earth's surface, low orbit, and geosynchronous orbit are a rather poor way of estimating the difficulty of getting from one place to another.

There are a number of different ways of expressing the difficulty of getting around in space, and often one has to analyze the flight plan of a particular mission in order to determine just how much effort is required. The most compact way of illustrating the peculiar geography of space is through a number which is usually written $(\delta)V$, pronounced "delta-vee." "Delta" (δ) is a Greek letter conventionally used by scientists and engineers to express the notion of the change in some quantity; the capital V

denotes velocity in this case. In order to make life easier for readers and typesetters, I'll spell the term out as delta-vee in what follows.

Because of different mission profiles, it's not possible to map out the inner solar system with different delta-vees in the same way that maps of the earth can be made as accurate as patience and money allow. What's the delta-vee to go to Mars? Well, it depends on just when the trip takes place, and whether the trip is to the Martian equator, to the poles, or somewhere in between. However, these differences are matters of detail, and don't affect one's general conclusions about how much rocket fuel—and how much it costs—to get to various destinations. Table 6 provides you with some delta-vees for various destinations of interest.[1]

Exploring Delta-Vee

Let's consider a trip to the surface of the moon. To get from the earth's surface to low orbit (here defined as an orbit 500 km above the earth's surface; the shuttle's orbit is a bit lower) requires a delta-vee of 9.2 km/sec. Reaching the moon's orbit only requires a delta-vee of 3.2 km/sec, and to reach the lunar surface you have to make another change of 2.4 km/sec, using your rocket to slow you down so you don't crash. From this perspective, the "distance" between low orbit and the moon's surface, which adds up to 5.6 km/sec, is less than the distance between the earth's surface and low earth orbit. These numbers assume that the astronaut or computer controlling the spacecraft doesn't waste any fuel correcting for mistakes.

The trip back from the moon to low earth orbit introduces another new wrinkle. When you approach a planet with an atmosphere, like the earth or Mars, you can in principle use the atmosphere to help you out via a technique called "aerobraking." A spacecraft's trajectory can carry it into the top of the earth's atmosphere, where friction will slow it down. It still takes 2.4 km/sec to get off the lunar surface, but use of aerobraking reduces the delta-vee from lunar orbit to low orbit to a small 0.7 km/sec, for a total of 3.1 km/sec for the entire trip.

Aerobraking is a new technology and hasn't been tried yet, though a closely related technology called "aerocapture" has been used since the very early days of the space program. In both cases, the spacecraft must approach the upper layers of a planet's atmosphere at a precisely calculated angle, with the spacecraft pointed in the right direction so that the heat shield can do its job. When the space shuttle lands, it loses virtually all of its orbital velocity and slows down so that it can come in on a

Table 6. Reaching Various Destinations

Trip	Delta-Vee (km/sec)	Trip Time
Earth surface → low earth orbit (500 km)	9.2	˜10 min
Low orbit → geosynchronous orbit	4.2	12 hours
Low earth orbit → lunar orbit	3.2	3 days
Low earth orbit → earth/moon L[1]	4.4	3 days
Lunar orbit → lunar surface	2.4	1 hour
Lunar surface → low orbit	3.1[a]	3 days
Low orbit → Phobos or Deimos	5.6[a]	270 days
Phobos/Deimos → low orbit	1.8[a]	270 days
Low orbit → Mars	4.8[a]	270 days
Mars → low orbit	5.7[a]	270 days
Low orbit → typical near-earth asteroid	4.5–5.5	≈1 year
Typical near-earth asteroid → low orbit	0.5	1 year
Low orbit → Asteroid 1982DB (most favorable case; January 2001)	4.5	210 days
1982DB → low orbit (most favorable case)	0.06[a] (!)	480 days
Round trips		
LEO to typical near-earth as teroid and back	5–6	
LEO to Phobos and back	7.4	
LEO to lunar surface and back	8.7	
LEO to Mars and back	10.5	
LEO ↔ Mars	10.5	

[a]Using aerobraking in the planet's atmosphere to slow the spacecraft down when it reaches the planet.

runway. The space capsules used in the Apollo program and in the current Soviet space program do the same thing, except that the goal is to slow down to the point where parachutes can be deployed. In the untried technique of aerobraking, the goal is to slow down somewhat less, so that the spacecraft can skip out of the terrestrial or Martian atmosphere and go back into orbit. There's no reason to suspect that aerobraking can't work, and in the rest of this book I will assume that it will.

The numbers in Table 6 can be translated into a new view of the solar system, where the value of delta-vee measures the distance between two points in the solar system. Dana Rotegard and Jeff Beddow, artists and space analysts at a small company in Minneapolis, have produced two illustrations which are reasonably good representations of how difficult it is to get around the inner solar system. The first illustration shows one-way delta-vees to various solar system destinations, and the second one includes round-trip delta-vees. For their own good reasons, Beddow and Rotegard have included a "typical" value of delta-vee for a near-earth asteroid; they are less optimistic about the asteroids as possible resource sites than I am. Our differences illustrate the difficulty of mapping the inner solar system precisely.

The earth–moon Lagrange (or Lagrangian) points are shown at the bottom of the first illustration. These points, found in any system of two orbiting bodies, are locations where an additional mass would be in equilibrium, with equal gravitational forces from the earth and the moon. L4 and L5 are stable Lagrange points, where a space station would tend to drift back to these locations after a small change in its position. While L1, L2, and L3 are unstable Lagrange points, it would only take a small rocket burn to push a station back to its correct location, and most of the time a station would remain in place. It takes less energy to put material from the moon into L2 than it takes to reach any of the other points.

Some thought about the numbers in Table 6 and the illustrations reveals some surprises. The most obvious has been referred to already, and is the reason to think about the concept of space resources in the first place: *The surface of the earth is a long way from virtually anywhere in space*. It's not that hard to get back to the surface of the earth from places in space because the atmosphere can slow you down. However, if we are to settle the solar system, most of our effort is going to be in getting people and cargo into space and not back down, and so the 9.2 km/sec needed to climb out of the earth's deep gravity well is a higher barrier than virtually any other barrier that we have to overcome.

A second feature of the table and illustrations is the small numbers associated with some of the smaller bodies in the solar system. Most people, when they think about space travel, think of the large, well-known bodies in the solar system like the moon and Mars. The problem with obtaining resources from these large bodies is that you have to use large quantities of rocket fuel to haul stuff out of their strong gravitational fields.

Compare Mars and its satellite Phobos, for instance. If you had a choice of returning cargo to low earth orbit from Mars or from Phobos, the choice is clear. Delta-vee from Mars is a large 5.8 km/sec, largely reflecting the effort needed to get off of the surface of Mars. Phobos is so small that an astronaut in a space suit could practically jump off its surface with leg power alone, and a mere 1.8 km/sec is required to get back to the earth. In both cases, these numbers assume you can use the earth's atmosphere for aerobraking.

Even more amazing are some small asteroids, which are in orbits that cross the earth. These tiny objects are so small that they don't even have

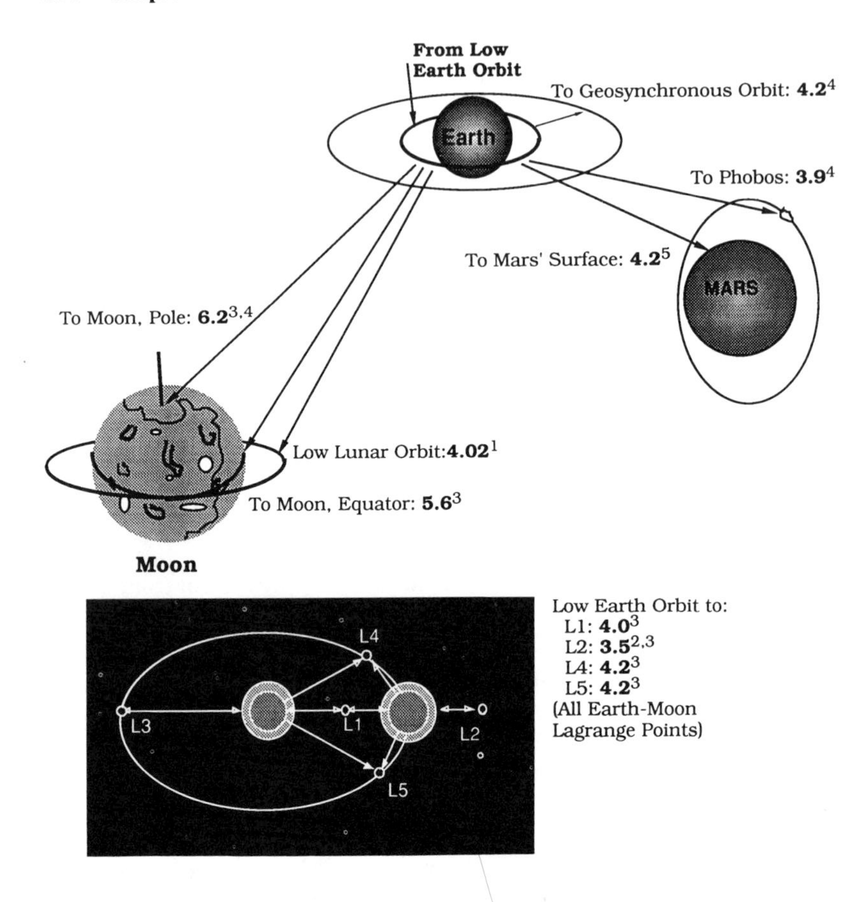

The top part of this illustration shows the one-way delta-vee required to reach various locations in space. (Delta-vee is the change in velocity required to get from one point in space to another. It is a good measure of economic access to various solar system bodies. Propellants and propulsion systems will change over time, but delta-vee is a function of mass. In this figure, delta-vee values are in km/sec.) The superscript numbers on the delta-vee values refer to various technical papers, cited in an article by Dana Rotegard.[12] The bottom part illustrates various stable points in the earth–moon system, where some visionaries have proposed to put a space station. (Courtesy of Jeff Beddow.)

names. Asteroid 1982DB (or DB 1982), listed in the table, is one of these; its numerical designation indicates that it was discovered in 1982 and the letters "DB" identify it by its order of discovery. If its orbit becomes better known, it will eventually get a name. It's a tiny little thing, only a

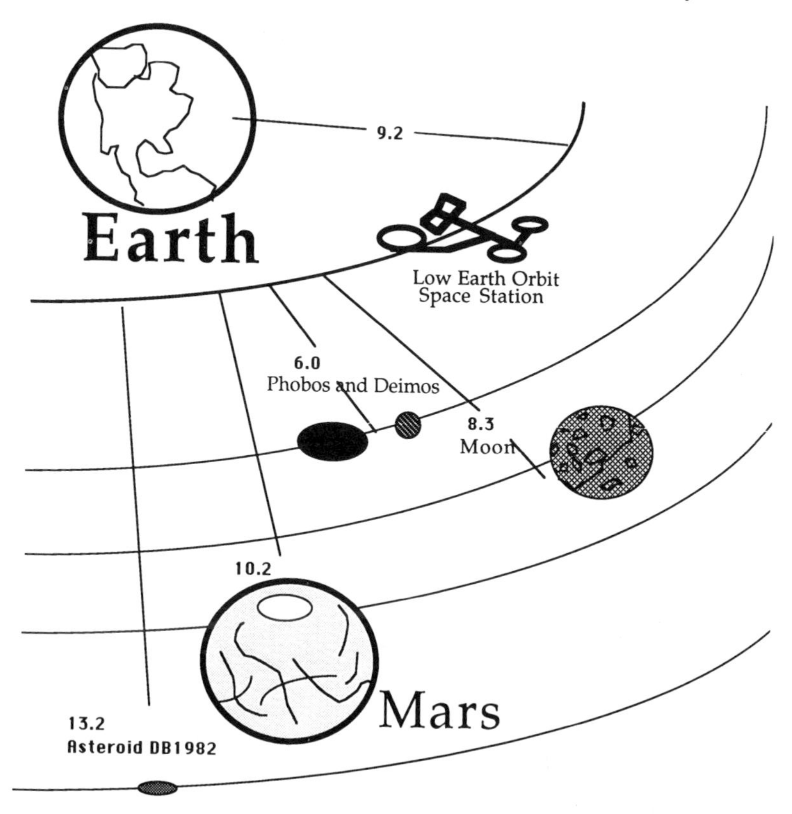

This illustration shows the round-trip delta-vees to various destinations in the inner solar system. The value for Asteroid DB1982 is for a launch at a "typical time" rather than a launch at an optimum time; under some circumstances, typical near-earth asteroids are as close as Phobos and Deimos. (Delta-vee values in km/sec.)

kilometer or so across, and landing on its surface requires very little rocket fuel. Because aerobraking can be used on the return trip, it's one of the easiest places in the solar system to get back from.

A potential difficulty with the asteroids, responsible for the different place of 1982DB in one of the illustrations and in Table 6, is that the delta-vee values needed to get back from them are critically dependent on the ability to schedule a trip at exactly the right times. Space flight experience so far has indicated that delayed launchings are far more common than on-time ones, which leads some people to feel skeptical about the exploitation of the asteroids. The very low delta-vee values only apply to on-time trips, and if the launch date is late, the value of delta-vee and the cost of the trip skyrockets.

Small objects like Phobos, Deimos, and the near-earth asteroids look particularly attractive for some purposes, where it's the return trip that matters. Suppose you're looking to space bodies as sources of water or shielding for a space station in low orbit. If the weight of the supplies you're importing is considerably greater than the weight of the transfer vehicle you're sending to get the supplies, then what really matters is the delta-vee for the return trip, since you can use aerobraking in the earth's atmosphere to slow you down. The low gravitational field of tiny Asteroid 1982DB makes it a very attractive source of supplies. If for some reason this particular asteroid won't do, there are many other asteroids in orbits that cross the earth. While the delta-vee required to come back from a typical near-earth asteroid is something like 0.2 km/sec, not as spectacularly low as it is for 1982DB, it's still very reasonable.

The difficulty of getting to the moon and Mars illustrates an amazing feature of these numbers. We have gotten both astronauts and machines to the moon and back, many times. Many people think that because the moon is only a quarter of a million miles away, and Mars is hundreds of millions of miles away along the usual path, Mars is impossibly more difficult to get to. However, the amount of delta-vee, related to the amount of rocket fuel, required to get to Mars is in fact *less* than that needed to get to the moon, largely because Mars has an atmosphere which you can use to help you land.

LIMITATIONS OF DELTA-VEE MAPS

Travel Times

The difficulties of space travel involve a number of trade-offs, and in some cases the delta-vee way of visualizing distance obscures the important role that these trade-offs play. For instance, saying that the delta-vee needed to reach Mars is less than that needed to get to the moon doesn't necessarily mean that it will be easier to send people to Mars than it was to send people to the moon, in spite of what some fervent advocates of a Mars expedition have stated or implied. If all you want to do is to send an unmanned spacecraft to Mars, it is indeed easier—it takes less rocket fuel, and the duration of the trip doesn't matter. Robot spacecraft don't drink water, don't eat food, and don't breathe air.

But the easiest way to get to Mars, with the lowest delta-vee, is to follow what's called a "Hohmann transfer" orbit, named for the person who figured it out, in which you follow an orbit which gently spirals around the sun and just reaches Mars with the right velocity. The difficulty with a Hohmann transfer orbit is that it takes a long time to get to your destination. If you want to go to Mars faster, you can do it, but it takes more fuel. For instance, you could follow a rocket trajectory which went straight away from the sun from earth to Mars. The delta-vee required for the speedy trajectory is huge, more than 50 km/sec.[2] Even though trip times could be cut to a month or so, such a path to Mars is clearly extravagant. However, a little boost to cut the trip time by a few months might seem worthwhile, and indeed such a boost has been widely discussed for the first human flight to Mars.

Trip times may matter when it comes to sending people to Mars. Whomever is paying for the mission has to supply the necessities of life for these people for years rather than the weeks that it takes to get to the moon. If we were to try to go to Mars with our present, American, throwaway technology, the tons of food and water you would need to send to Mars would more than make up for the tons of rocket fuel you save by sending people to Mars instead of to the moon. It does get rather complicated.

However, when you take the long view (as this book does), the delta-vee way of thinking about distances makes sense, even in the context of sending humans from one place to another in the inner solar system. When a closed life-support system like that outlined in Chapter 7 becomes available, travel times will be less important. The cost, or difficulty, of a journey may well be simply expressed by how much rocket fuel is needed to get from here to there, and delta-vee is a good number to think about to visualize the limitations imposed by rocketry alone.

Launch Windows

Perhaps the most debatable aspect of using delta-vee to describe the difficulty of getting around the solar system is that some destinations are considerably more accessible than others. An astronaut can only use a Hohmann transfer orbit to go from one place to another at certain times, when the two planets line up. These times when the delta-vee requirements are minimal are called "launch windows" because they open up for a while and then close again. The launch windows for the Mars expedition, and for missions to near-earth asteroids, open at approximately two year intervals.

Launch windows are, at best, open for a month; before and after that, the required values of delta-vee skyrocket.

Launch-window limitations for a return trip can be quite severe, especially if mission planners insist on the absolutely lowest possible value of delta-vee. The minimum energy path to Mars calls for a 9-month Hohmann transfer to Mars, an 18-month stay, and a 9-month return trip. Missions to asteroids depend on the particular asteroid, but in general are subject to similar constraints. Asteroid advocates John and Ruth Lewis point out that, in contrast to Mars, there are many accessible asteroids, and that at any particular time the launch window is open to some asteroid or another.

One advantage the moon has as a destination is that the launch window is always open. Because the earth is at the center of the moon's orbit, the minimum-energy, Hohmann transfer ellipse from low earth orbit to the surface of the moon is accessible each time the spacecraft circles the earth, or once every 90 minutes. Similar considerations apply to the return trip.

Complications

Because space flight is complicated, one simple number can't tell the whole story of how hard it is to get from here to there. For any particular journey, there are a number of additional factors which can increase the delta-vee that's required. Consider the trip to Mars as an example. The orbit of Mars around the sun is not exactly circular, and sometimes Mars is closer to earth than at other times. If you can plan the trip so that you land on Mars when it's closest to the earth's orbit, and to the sun, then delta-vee will be less. However, such opportunities only occur at intervals of decades; the 1992 launch opportunity is the most favorable that will occur for a while, much more favorable than the 1996 or 1998 ones, for instance. What value of delta-vee should be used in a map—the most favorable one, or the average one? Other questions introduce similar complications. Is "low earth orbit" at an altitude of 250 km or 500 km? Does "landing on Mars" mean "at the equator"?

In addition, the relative values of the delta-vees understate the difference between the low delta-vee and high delta-vee destinations. The amount of rocket fuel you need in order to get somewhere goes up exponentially with increasing delta-vee. The way that rockets work means that you have to use a lot of fuel at the beginning, accelerating the remaining fuel and the payload, not just the payload alone. For example, it takes 3.75 million pounds of hydrogen and oxygen to accelerate the space shuttle to its orbital velocity of about 9 km/sec. If you were to strap on a second external tank to try to give the shuttle twice as much fuel, then you could only boost it to a speed of 11 km/sec.[3]

Perhaps the simple delta-vee number isn't enough to describe the difficulties of any particular trip. However, in order to think about where we want to go in the solar system, a simple number which can be used as the basis of maps is needed. So, for the purposes of this book, I will stick with delta-vee, warts and all, as the best representation of distance.

ADVANCED PROPULSION SCHEMES

Rockets which can reach space have been around for over 50 years, and the basic idea has not really changed since some unnamed Chinese warrior started shooting them at his enemies in the 1200s. You burn rocket fuel, usually supplying oxygen to it by mixing the fuel with an oxidizer, and design a rocket nozzle so that the hot gases from the burning fuel get shot out the back and push the rocket forward. The most serious limitation on rocket performance is that you can't shoot exhaust gases out the back anywhere near as fast as you want to push the rocket forward. As a result, you need to use enormous amounts of fuel in order to push a tiny payload into space. The space shuttle uses roughly 20 pounds of fuel to put each pound of payload into earth orbit.

A real difficulty with an interplanetary mission like a manned Mars landing is that part of the starting payload is simply the fuel that you need in order to get back. As a result, the ratio of fuel to payload multiplies to enormous values. For instance, the analysis of one particular Mars trip called for 163 million pounds of fuel to lift a 240,000-pound payload with 11 crew members in it to Mars and back. This works out to about 700 pounds of fuel for each pound of payload.[4] Of course there are many ways to accomplish a Mars mission, and since much of the machinery isn't designed yet, numbers like these should only be considered illustrative. I bury this calculation in a footnote to show that the very rough numbers which I cite in the text are not that far off what may become reality, *if* we use chemical rockets and nothing but chemical rockets to send people to Mars.

The numbers suggest that some modest improvements in rocket capabilities could pay enormous dividends; the cost of a Mars mission at present is dominated by the cost of the fuel that is needed. Engineers have tried to to make rockets better by using new materials in the engines. That way, the engines operate at higher temperatures and you get some modest increases in exhaust velocity and thus rocket performance. But any big change will require a radically different technology. It would seem presumptuous to try to predict technological advances into the 21st century, a time period which is longer than the time that's passed since the introduction of the automobile. Yet there are some new propulsion schemes with great promise, which can dramatically reduce the amount of fuel required per pound of payload. Some work was done on these in the 1960s and 1970s, but they are now forgotten, nowhere to be found in NASA's planning documents.[5] Neither one is included in Project Pathfinder, NASA's current effort to develop the technologies needed to go to Mars.[6] While a single trip to Mars could probably be done without them, successful development of these technologies could help enormously in the settlement of the inner solar system. If Project Pathfinder is to reach beyond the initial trip to Mars, surely some work on advanced propulsion technologies is called for.

Electrical Propulsion Systems

Electricity can accelerate charged particles to speeds which are much higher than can be attained in a conventional chemical rocket. The general concept of electrical propulsion is relatively simple. Suppose, for example, a normal battery were to be used to power a rocket. Individual atoms are stripped of one or more outer electrons and become electrically charged. These charged atoms (called "ions") are attracted to the positive terminal of the battery. Connect the positive terminal of the battery to the right kind of grid, far enough away from where the ions are generated, and they can pick up a lot of speed on the way to the positively charged grid. Exhaust velocities of roughly 30 km/sec, six times higher than that of chemical rockets, are possible.[7]

Electrical propulsion is rather different from chemical rockets in that the ions leaking out the back of the rocket produce a very weak force. A single ion thruster pushes on the spacecraft with a force of only a few pounds, and so someone in the spacecraft would be unable to tell whether the ion engine was actually working or not. However, these engines can be turned on continuously, for the entire mission. The electricity required for this type of rocket engine can be supplied either by a nuclear reactor or by solar panels.

Electrical propulsion schemes currently under design use some rather lethal materials. A favorite candidate for the ion to be accelerated is mercury, because it is heavy and easily ionized. Heavy atoms make ion engines work better because one atom can give a greater push to the spacecraft. The most intensively developed concepts, and the ones discussed most frequently at the present time, are those where a nuclear reactor provides the electric power. Nuclear reactors and mercury are both quite hazardous.

A nonnuclear alternative is to power the ion engines with solar cell arrays. This alternative is quite advantageous in that the ultimate energy source is sunlight, which weighs nothing and doesn't have to be carried up from the earth. Such a system was studied in some detail as a way of sending a spacecraft to Halley's Comet. When the comet mission collapsed in the budgetary crises of the late 1970s, NASA's ion engine studies came to an abrupt halt as there was no longer any mission on NASA's plate which required these new engines.[8]

The future of electrical propulsion schemes is not clear at this time. NASA had spent roughly $3 billion on various nuclear-powered propulsion schemes between 1959 and 1973, when the program was canceled. When projects like these are canceled, it's not that easy to start them up where they were left off. Many of the designs are located in the offices of various engineering companies around the country, and some of these companies may not exist any more. Designs may depend on components which are no longer manufactured. Most important is that much of the knowhow resides in the minds of the engineers who worked on the project, and many of these people are now either retired or else fully occupied on some other project.[9]

Solar Sailing

From an esthetic point of view, solar sailing is by far the best way to get around the solar system. Virtually everyone whom I've met, sailors and nonsailors alike, is captivated by the esthetic beauty of sailboats. Huge expanses of fabric, pushed out by the breeze into soft, billowing curves, force a graceful boat to cut through the water toward its destination. This is harnessing the forces of nature at its best. Columbus didn't need to carry a load of smelly fuel along with him; he used the power of the wind to cross the ocean.

The very notion of solar sailing creates an image of something similar in space. Graceful, silent spacecraft simply catch the sunbeams and practically float from one planet to the next. A tiny habitation module, tens of meters across, is dwarfed by a huge, thin sail half a kilometer on the side.

But still, the pilots in this habitation module can control the huge expanse of solar sail, using the forces of nature to say farewell to the powerful blast of chemical rockets.

Just as a real sailboat falls short of the romantic picture of terrestrial sailing, so is there a huge difference between the esthetics and the reality of solar sailing. The gently curved terrestrial sails which you always see in pictures can hang limp and impotent if there is no wind or can turn into uncontrollable, flapping monsters in a stiff gale. Similarly, the idealized picture of solar sailing produced in some early science fiction novels crashes to the ground in the face of ugly facts.

Solar sailing uses the energy in sunlight to change the velocity of a spacecraft. Despite the name of the technique, the solar wind is not involved in solar sailing at all. A solar sail is nothing more than a huge, ultrathin mirror which sunlight bounces off of. The light from the sun comes in a stream of tiny particles called "photons." When these photons bounce off a mirror, they exert a small force on a mirror, just as air molecules deflected by a cloth sail exert a force on that sail.

The basic principles of solar sailing are quite simple. Easiest is using the sun to force the vehicle to spiral outward. Control of the solar sail is exercised by changing the angle of the sail with respect to the direction of the sunlight which hits the sail. If the astronaut controlling the sail angles it so that the rays of sunlight, or photons from the sun, bounce off it backward, relative to the sail's direction of motion, the force of these photons on the sail will push it forward, increasing its orbital velocity and pushing it away from the sun.

It's a bit more subtle to see how a solar sail can allow you to travel inward, toward the sun. The sail is angled so that the photons which bounce off of the sail travel forward, in the same direction that your spacecraft is moving. The reaction force on the sail is, now, in the backward direction, so the solar sail slows down its orbital motion. A slower orbit will cause the sail to travel in an elliptical orbit, closing in on the sun. If you deployed the sail forever, your path would be a giant death spiral which would carry the sail into the inner solar system and fiery destruction. An astronaut can control what the sail does and stop the death spiral when the final destination is reached.

Some realism—in other words some numbers—dampens the romance of solar sailing a bit. A $4-million study done at the Jet Propulsion Lab (JPL) in the late 1970s has produced some nice concepts, though no hardware was ever built or tested. The goal of this study was to design a spacecraft which would do a mission to Halley's Comet in 1986. The solar sailing concept, like the Halley's Comet mission, fell victim to budgetary pressures from the space shuttle program.

The pressure of sunlight is pretty weak. Realistic designs indicate that a 1000-kilogram spacecraft could be pushed around the solar system by a sail with an area of a third of a square mile (1 square kilometer). Such a sail would have an area of 200 acres; it could be a square a little over half a mile on a side. The Apollo command module, holding three people, would require a sail with four times the area. These are huge, gossamer spacecraft. Solar sailing forces a designer to think big.[10]

How does one control a 200-acre sail? To work, solar sails have to be extremely lightweight, and thin. Household plastic wrap is about four times too thick. Yet these sails have to be controlled, and anyone who has fought with plastic wrap knows how hard it is. Imagine trying to get 200 acres of the stuff to stay flat!

The tilt of the sail relative to sunlight is critical, so the sail has to be rigid. One possibility, of course, is to stretch the sail on rigid spars. Sails like this are called "square" sails, though they don't have to be square. The trouble is that the weight of the spars is then added to the weight of the sail and it doesn't work as well. Another possibility is a spinning disk sail, where centrifugal force keeps the sail rigid. Detailed analyses show that the spinning disk sail might be difficult to control. A third concept is the "heliogyro," an extension of helicopter design on a grand scale. Here, the sail elements are long, skinny "blades" that are spun from a central axis. To get a total reflecting area of something like a square kilometer, the blades have to be very long. Do the arithmetic, and you find that a 20-blade heliogyro with blades 10 meters across and an area of 1 square kilometer would have blades that were 5 kilometers long. You have to think big to think about solar sails.

Currently, the only work on solar sails is being privately financed by foundations like the World Space Foundation in the United States and the Union Pour la Promotion de la Propulsion Photonique in France. These foundations exist to keep the dream of solar sailing alive. Louis Friedman, who directed the JPL's solar sailing study, argues that solar sailing has been strongly supported by the public at large and by bright young engineers, but resisted by bureaucrats.[11] We're barely into the "concept formulation" stage in solar sailing, and before anyone starts counting on solar sails working we certainly need a proof of concept test in which somebody actually tries to deploy a solar sail in orbit, even a small one, to see whether it can be unrolled in a stable way. Many barriers separate the dream of solar sailing from a possible reality of regattas in the inner solar system.

* * *

The few hundred miles which separate the earth's surface from low earth orbit might seem microscopic in comparsion with the millions of miles that separate the earth from other planetary bodies, large and small. However, because the earth's gravity is so strong, these few hundred miles loom as an almost unbridgeable chasm which makes access to space difficult. Indeed, the peculiar nature of space flight makes it necessary to think about distances in a new way. From the point of view of the rocket fuel involved, the surface of Mars's satellite Phobos is nearer to low earth orbit than the surface of the moon.

Chemical rockets could certainly propel us around the inner solar system, but they seem clumsy and inefficient compared to some promising if unproven propulsion technologies like solar sailing and solar- or nuclear-electrical propulsion schemes. The question remains, though, whether any of the planets or satellites in the solar system do in fact contain the resources which space stations need. This question is the focus of the remaining chapters in this section.

CHAPTER 9

The Moon

A Worthless Piece of Real Estate?

The Luna City Times

October 4, 2020

Transmissions from Polar Rover 13 indicate that an icy deposit in a permanently shaded lunar crater has at last been discovered. These long sought deposits of lunar polar ice have survived since the beginning of lunar history, 4.5 billion years ago. At last, human settlements on the moon may be freed from our dependence on Earth for that critical element, hydrogen. We can make oxygen from lunar rocks, but making water requires us to import hydrogen at great expense from our home world. Perhaps this is the critical discovery which will lead to lunar independence. . . .

The next shipment of shielding to Space Station 7 will be sent off next Sunday. This ton of bricks, black glass made from lunar soil, will be used as radiation shielding for this space station.

Are these fictional dispatches from a lunar newspaper bound to remain forever in the world of science fiction, or could they possibly become science fact someday? The moon is the natural target which everyone thinks of when considering places beyond the earth where human beings can establish an outpost. In terms of miles and trip time, it's by far the closest extraterrestrial body. While there has always been some interest in exploring the moon, other worlds now seem far more interesting than this collection of craters. In the last ten years, exploration of the moon has become less important than exploiting our satellite as a source of space resources.

The moon is also an inviting target because we know so much about it. Astronauts have already walked on the lunar surface, collected hundreds of pounds of rocks, and to some extent mapped it. It's familiar; everyone recognizes the moon, and can identify where we're going. The potentially critical role of the moon in the 21st century space program is somewhat ironic, because it was far from the astronomical mainstream for much of the 20th century. The surface of the moon contains practically nothing but impact craters, and the simple act of counting and naming craters became quite boring and uninsightful after a while. Virtually no one specialized in lunar studies through the 1950s. Pioneering lunar scientist Harold Urey, writing in his memoirs, cynically referred to the moon as a "worthless piece of real estate," describing not his own views but rather the views which prevailed in the scientific community at the time he joined with NASA to urge a lunar landing in the early 1960s.

INTRODUCING THE MOON

When Galileo first turned his telescope on the moon in 1610, he realized that this pale, luminescent ball with faint markings on its surface was in fact a rather mountainous world with craters, light areas, and dark areas. As seen in the illustration on page 32, the most ubiquitous component of the lunar landscape is the circular crater. (The craters near the edge of a picture of the moon may look elliptical, but that's because the moon is round and we are looking at these craters on a slanted line of sight.) It turns out that even the large light and dark areas, which create the appearance of a face in the moon, are made up of large, overlapping circles.

Galileo had no idea of what the moon was really like. In his day, the prevailing opinion was that the earth was the center of the Universe and that the moon and planets were tiny light bulbs which the Almighty put around the earth to light up the sky at night and amuse us. He named these large circular areas *mare*, which means "seas" in the Latin which Galileo used for most of his scientific writing. The large circular area at the bottom center of the moon in the illustration was called *Mare Imbrium* ("Sea of Rains"). The name could not be more inappropriate on the moon. This Sea of Rains was never filled with water, and it never rained there (or anywhere else on the moon, for that matter). Such a name might make sense on Mars, where at least there's evidence for water flow if not rain. But if rains didn't make these major dark areas, what did?

ORIGIN OF THE LUNAR SURFACE FEATURES

When the moon first formed it was quite hot, hot enough to be molten from center to surface. As it cooled, chemicals such as iron and magnesium silicates which solidified first sunk deep into the moon's surface, depleting the surface layers in particular elements. Because the moon is so much smaller than the earth, it cooled off quite quickly. Liquid rock, or magma, was close to the surface only in the first billion years of its evolution. At one point during this time, a number of 50-mile objects whizzing through interplanetary space hit the moon in various places. These impacts produced the roughly circular dark-colored areas which Galileo called *Mare* ("seas") and which make up the "features" of the man in the moon.

An intuitive, but incorrect, picture of how a crater forms on the moon is that the incoming body simply digs a hole in the lunar surface. The true picture is a bit more complicated, but explains why all of the craters on the moon are circular. The incoming object has a great deal of energy simply because it's moving at a speed which is, for most reasonable orbits, about 30 km/sec or 175,000 miles per hour. The 50-mile boulder which carved out *Mare Imbrium* had more energy than a hundred thousand Hiroshima-type atomic bombs. This energy is suddenly released as the incoming boulder suddenly stops moving because it slams into the lunar surface, causing an explosion. The result of an impact is, then, like setting off a bomb some distance below the lunar surface.

As long as the part of the lunar surface above the explosion is reasonably uniform, this explosion will hurl lunar debris out evenly in all directions. Directly at the impact point, and for some distance around, the moving moonstuff will be tossed away, and will land on the lunar surface some distance away. The result of this explosion is a circular crater, just like a bomb crater.

If the impact occurred early enough in lunar evolution, as the *Mare Imbrium* impact did, the cracks created by this megaexplosion reached down far enough so that at the bottom of the cracked layer some molten lunar rock, magma, existed. This magma, being under pressure, seeped upward through the cracks and oozed onto the lunar surface. The magma also had a different composition from the lunar surface, since it contained more of the iron and magnesium which crystallized first and fell into the lunar interior. When it oozed out onto the lunar surface and solidified, it was darker than the material of the surrounding region, explaining the color contrast.

The dark areas on the moon are, then, filled-in impact basins which were all created by bombardment less than a billion years after the moon's origin. We don't understand just why the intense bombardment ceased; perhaps impacts on the moon and on the other planets used up all of the orbiting debris in the inner solar system. Since then, the only thing that has happened on the moon is the occasional impact of another rock. Some of these impacts can create quite spectacular craters like the bright-colored one in the right center of the picture of the moon. Because the moon's surface is quite old, all of the craters which were ever created in its 4.5 billion year history are preserved on its surface, resulting in a pockmarked landscape which shows its age.

Comparison with Earth

The earth's surface, in contrast, is quite different. Earth has irregularly shaped continents, high mountain ranges, volcanoes, rivers, flat plains, and very few impact craters. Why?

One reason is that the earth has an atmosphere and a water cycle, and the moon does not. Rainfall erodes high mountain ranges and washes away topographic features, washing the sediment to the sea. When lakes and seas evaporate, large flat plains are created. Alternate freezing and thawing of water is a powerful way of fragmenting solid rocks and creating soil. But erosion itself can't explain everything; what created the mountain ranges which water then erodes away and flattens?

In the last 50 years, the idea of continental drift or, more accurately, plate tectonics has become the conventional wisdom in geology. The German geologist Alfred Wegener was struck by the similarity of the eastern coastline of North and South America and the western coastline of Europe and Africa. He proposed that these continents actually drifted apart. This radical idea was confirmed in the 1960s when geologists detected evidence of sea floor spreading in the central Atlantic Ocean.

The terrestrial landscape is shaped, then, by drifting continents. When two continental landmasses bang into each other, land is squished up against land, producing folded mountain ranges like the Himalayas and the Rockies. When a continent drifts over an ocean, ocean bottom plunges beneath the continent. These gigantic collisions produce earthquakes and volcanoes; the reason that many volcanoes are concentrated around the rim of the Pacific Ocean is that oceanic crust is being subducted beneath continents on many Pacific Ocean shorelines.

The reason that few impact craters survive on earth is that the earth's crust is constantly being sculpted and resculpted by these powerful geological processes. Impact craters are formed on the earth at roughly the same rate they are formed on the moon, but the combined forces of plate tectonics and weathering—geologists' names for continental drift and erosion—erase them. Maybe there was the equivalent of *Mare Imbrium* on earth 4 billion years ago, but it would have disappeared because it was scrunched together to make a mountain range, buried in oceanic or lake sediment, or possibly simply eroded away. In parts of the earth's crust which have survived for a long time in an undisturbed way, like most of Canada, there are a number of impact craters around, though they are a bit harder to distinguish because they have weathered away to near oblivion.

LUNAR RESOURCES

The exploitation of earth resources, or mining, is carried out in a way which may well prove to be applicable only on our planet. All of the geological excitement of plate tectonics, volcanism, and weathering has an important practical consequence. In ways which are often not completely understood, these processes produce local concentrations of some economically important minerals. In some cases, all an entrepreneur needs to know is just where to dig; valuable materials like gold and diamonds are simply there for the taking once you find them. In other cases, the elements of interest are still quite concentrated in ores; a mining company can simply pick out the right kind of rocks and then use simple chemical processes to extract the good stuff from the mine tailings. There's plenty of oxygen available if it's needed for these chemical processes, and usually (though not always) water can be readily obtained too.

The moon, though, is different. Most importantly, we can conceive of few geological processes which would operate on the moon to concentrate particular types of minerals in particular areas. The results from the various lunar landing expeditions bear this out; while there are different types of moon rocks, the same major rock types were found at the six different Apollo landing sites. These sites were all in the lunar *mare*, the large dark areas, and so the dominant local rock type was the dark-colored rock which geologists call "basalt" characteristic of the *mare* regions. The other major lunar rock type, the light-colored one from the "highland" regions between the *mare*, contains much less titanium and iron, and more aluminum.

The moon, then, does not appear to contain the elements of a lunar gold rush. What we know of its geology suggests that the same major rock types are likely to be found at just about any location on the moon's surface. It's true that we have detailed samples from only a few locations on the lunar surface, namely, the landing sites where astronauts and automated landers picked up rocks and brought them back to earth. We also have some familiarity with the equatorial regions of the moon, mapped by instruments on orbiting Apollo spacecrafts. For the rest of the lunar surface, we can only look at photographs and make guesses.

Lunar Soil

The surface of the moon is covered almost uniformly with a lunar "soil" which is, however, quite different from terrestrial soil. The difference is so great that lunar scientists use a different word, "regolith," to describe it. Terrestrial soil is the result of a number of chemical and biological processes which have broken rocks into small pieces, enriched them with organic debris, and transported them over great distances. Depending on the underlying bedrock, climate, types of plants or trees on the surface layer, and the role of glaciers and rivers in soil transport, soils can vary greatly in their characteristics.[1] The lunar soil, in contrast, has been produced by a succession of impacts of particles big and small which have ground the lunar bedrock to a fine powder.

The lunar soil can, in principle, be quite useful for an expedition which is planning to exploit lunar resources, since it is already ground up and is easier to shape, sort, or mine than the rocks. It consists of a very fine powder, somewhat like black talcum powder. Most of the grains are mashed-up lunar rocks, though sometimes the force of an impact melts the lunar material and produces a tiny glass bead. We don't know as much about the lunar regolith as we would like. In 1969, no one realized that the lunar soil would be much more exploitable than the lunar rocks, and so the moon walkers spent lots of time picking up rocks and little time on dirt. Furthermore, removing soil from the moon changes it in important ways. For instance, its texture is like that of activated charcoal, and gases like terrestrial water vapor can stick to its surface.

While, in principle, it would be nice to do a number of experiments with the lunar soil, the amount available for space resource experimentation is quite limited. Only about 15 pounds of lunar soil left over in the rock collection bags is, apparently, available to scientists who wish to use it to simulate lunar resource exploitation. We know enough about lunar soil so that it's possible to use the right kind of terrestrial materials for some experiments.[2]

An important question which influences our decision on whether or not it is worthwhile to go back to the moon is just what can be done with the lunar rocks and soil. The earlier discussion of just what space settlements will need placed great emphasis on the role of water. Most of the moon is a very dry place. Any water which might have been present initially would have evaporated, and the moon's gravity is so weak that it cannot hold onto an atmosphere for any appreciable time. This severe limitation must be considered when contemplating the economic value of moon dust and moon rocks.

USE OF LUNAR RESOURCES

Radiation Shielding

Any space settlement, and particularly any space settlement outside of low earth orbit, must be shielded from the high-speed particles which the sun spits out occasionally. Any material will do for the shielding; while there are minor differences in the effectiveness of different elements as radiation shields, these differences are small enough that material availability is the overriding factor. The amount of shielding needed depends on how cautious you are. Several yards of average density rock will produce an amount of shielding equivalent to the earth's atmosphere, but this much may be unnecessary. Scientists investigating lunar base design suggest that about 6 feet of compressed lunar soil should suffice for most conditions. A permanent base should include a "storm cellar" with about twice as much lunar material, which could handle the extreme conditions of solar flares. If babies are to be born on the moon, the existence of a storm cellar is particularly critical, since fetuses are more sensitive to radiation than mature people.[3]

If we are to build a lunar base, we will use lunar material for shielding, either by piling stuff on top of a structure built on the lunar surface or by digging into the lunar surface. No one designing a lunar base would be stupid enough to use shielding lugged up from earth, at a cost of $10,000 per pound, where the lunar regolith was available for free.[4] All you would need is some kind of a backhoe to push the stuff around.

The scenario for lunar base construction seems quite clear. The construction of a lunar base begins with an automated mission, or several missions, landing on the lunar surface and deploying a base module. The robot backhoe digs a trench, and another automated piece of equipment lowers the base module or modules into place. In other scenarios, the base is put in place on the lunar surface and the backhoe or dragline buries it.

Connecting the modules before they are buried might be done by remotely operated pieces of equipment under terrestrial control, or it might require a visit by a human construction crew. Once the base is buried, with solar power cells remaining on the lunar surface, the first human inhabitants can occupy the facility.[5]

However, if the lunar base is to justify its existence in the long term, it must produce some material for export. A widely suggested possibility is that shielding for rocket ships which make the Mars run could be obtained more economically from the surface of the moon than from the surface of the earth. The delta-vee numbers do support this idea, but you then have to ask just what you would do with the crumbly regolith in order to transform it into solid material which could be used either in a space station beyond low orbit or on the Mars run.

Preliminary studies for lunar bases have come up with some interesting ideas. If the regolith is heated to high temperatures, it will form glass bricks. These bricks will be black because of the very dark color of lunar materials, especially in the *mare*. Asteroid scientist and science writer John and Ruth Lewis point out that finding uses for dark glass bricks isn't easy.[6] Glass has a nasty tendency to break under stress; would the Mars craft's radiation shield shatter when its rocket engine was first turned on? The glass bricks would need to be fastened together with some kind of a sealant; putty, used in most houses, would probably become hard and brittle in the vacuum of space. These may not be fatal problems. A glass house in space, however illogical it might seem, might even work.

Another thought—that lunar soils could make good concrete—sounds appealing, at first. Many terrestrial structures are made out of concrete, for good reason. When concrete is slushy, a builder can mold it into virtually any shape you want, ranging from suspension bridge supports to the exotic, sail-like shapes of the Sydney Opera House in Australia. Once it hardens, it is quite strong and is not damaged by rain. Construction workers can still drill into it and put bolts in if there's a need to attach anything to it. Repair is easy.

Experiments with lunar soils show that they do indeed make good concrete. They contain enough of what concerete experts call "cementitious materials"—mostly calcium oxide, but including oxides of silicon and aluminum.[7] To make concrete, a builder puts ground up cementitious materials (called, not surprisingly, "cement"), rock, and water together to make a slurry. The composition of the rock really doesn't matter, and there's plenty of rock available on the moon. This slurry is then poured into molds of the appropriate shape and left to harden. It would then be a simple matter for workers at a lunar base to export radiation shields, in pieces, for use by terrestrial space stations or spaceships making the Mars run.

There is, however, a slightly sticky problem. On the earth, it's easy to get water to make cement, but on the moon the availability of water is a very serious problem. The only water in the samples picked up by the Apollo astronauts was small bits of water which stuck onto the fine grains in the lunar soils, and even this was present at the low level of tens of parts per million. Someone using lunar concrete would have to obtain the water from somewhere, either by finding it somewhere on the moon or else by importing it, at great expense, from the earth. There are a number of ways around the water problem which might be worth exploring.

Lunar Ices at the Poles?

The lunar polar regions have intrigued scientists for quite some time. The moon rotates about an axis which is almost precisely perpendicular to the plane of the earth's orbit around the sun; its axial tilt is a mere degree and a half compared to the earth's 23 degrees. As a result, sunlight is always nearly horizontal at the poles; there is no lunar equivalent of the long polar day or the long polar night. Orbiter photos from the 1960s show that the lunar landscape at the poles is quite rugged, and that there are crater bottoms which are at the present time permanently in shadow. These crater bottoms must be very cold. Ice evaporates quite quickly in the lunar vacuum if temperatures are high, but the evaporation rate plummets as temperatures fall toward the molecular inactivity of absolute zero. A number of scientists have speculated that under such conditions trapped ice could survive over the 4.5 billion years of lunar history.[8]

Whether icy materials exist at the poles is an important open question. Asteroid enthusiasts John and Ruth Lewis again provide a skeptical perspective to balance the lunar enthusiasts. Impact craters are found all over the moon, including at the lunar poles. Wouldn't the force of impact splash the ice out onto the lunar surface, from whence it would evaporate immediately and dissipate into outer space? Is the rotation of the moon really so stable that the bottoms of polar craters have never ever seen the sun?[9]

Obviously, the only way to answer definitively whether there is water in lunar polar craters is to dig there. Remotely operated landers could do just that, at considerable expense. However, remote sensing technology could determine whether ice at the lunar poles was at least possible. Such remote sensors have been proposed for a number of years. Gamma ray spectrometers can map the composition of rocks over the surface of the moon by detecting, from lunar orbit, the characteristic emissions from various elements. Infrared spectrometers could measure the temperature at crater bottoms and provide additional compositional information.[10]

A mission with all these instruments, called Lunar Polar Orbiter, has been on NASA's space agenda ever since the Apollo program ended in the early 1970s. It received relatively low priority treatment from a strictly scientific viewpoint because the moon, consisting largely of an old, cratered landscape, is geologically less interesting than some other planets. Nonscientific Federal policymakers were even less enthusiastic: "Why do you want to go to the moon? You've been there already."

Because the importance of space resources had not been recognized until recently, advocates of a return to the moon had no ready answers to such skeptical questions. But now that we recognize the importance of lunar resources, the Lunar Polar Orbiter mission may receive higher priority treatment. When and if it flies, it will, no doubt, not provide absolutely definitive answers. Remotely operated lunar rovers could visit the most promising sites and return samples from the bottom of lunar polar craters to terrestrial laboratories.

The Soviet Union may execute the important Lunar Polar Orbiter mission before we do. Soviet space scientists, in the first sign of space *glasnost*, provided a detailed announcement of its future space plans at an American conference in March 1985. These plans included a lunar polar orbiter mission to be launched in 1989 or 1990.[11] Two and a half years later, the Soviet space agency hosted a huge "birthday party," celebrating the 30th anniversary of the Sputnik launch by inviting space scientists from around the world to Moscow. At the celebration, they announced that the launch date for the polar orbiter had slipped to 1993.[12]

The Hydrogen Problem

Even if there is no water on the moon, it's possible that one component of water, which contains much of its mass, could be derived from lunar surface materials. It turns out that by far the most common element in the lunar crust is oxygen. Lunar rocks, like terrestrial rocks, consist largely of oxygen combined with other common elements like silicon and iron. For instance, a pure silicon-oxygen compound, SiO_2, is simply quartz. Most chemicals in the lunar rocks are in the form of silicates; feldspar, a common rock, is potassium aluminum silicate [chemical formula ($KAlSi_3O_8$)], with calcium or sodium occasionally appearing in place of aluminum. Oxygen is quite common in such compounds; 40–45 % of the weight of a sample of lunar regolith is oxygen.

What's intriguing is that most of the weight of water is in the form of oxygen. While water, H_2O, contains two atoms of hydrogen for every atom of oxygen, oxygen atoms are 16 times heavier than hydrogen atoms. As a result, then, 16/18ths or 89% of the weight of water is oxygen.

The abundance of oxygen on the moon and the high weight content of oxygen in water suggest that lunar oxygen might be useful, even by itself with no hydrogen. A lunar base could manufacture liquid oxygen for use in refueling spacecraft shuttling to and from the lunar base. While the spacecraft would still have to carry liquid hydrogen (or some other fuel) for the return trip, there might be a significant economic payback depending on the cost of producing oxygen on the moon from lunar soil.

Imported hydrogen could also be combined with lunar oxygen to produce water. In this scenario, lunarians would only need to import 11% of the weight of the water which was ultimately produced. The water could be used to make concrete, or to support life in the lunar base. Lunar oxygen could even supply a Martian spaceship or terrestrial space station, enabling the station to only import the hydrogen from the earth.

Such a scenario, though, is not terribly attractive from several perspectives. Hydrogen import would reduce the cost of water on the moon to a "mere" $2000 per pound along with whatever it cost to produce the lunar oxygen. At that price, lunar concrete is not going to be a terribly attractive material for a space shield on the Martian spaceship. Furthermore, the need to import large quantities of hydrogen would make the lunar base considerably more dependent on supply ships from earth. Human history suggests that colonies that thrive are those which quickly become independent from their homelands. An umbilical cord consisting of massive hydrogen shipments would tie the fate of a lunar base too closely to decisions made on earth.

* * *

The idea of bases on the moon, then, is not purely technological fantasy, though important questions need to be answered in positive ways before it becomes reality. The key question is whether there is any source of water—or hydrogen—on the moon. Without water, a lunar base could only export crudely processed lunar materials which might possibly be useful in shielding a space station or spacecraft making the Mars run. The utility of glass bricks remains to be demonstrated.

With water, a base can become truly self-supporting. Water mixed with lunar soil can make concrete, a very useful structural material. The water itself is important as a resource for space settlements anywhere in the solar system, and the comparatively weak lunar gravity field makes the moon a much better water resource than the earth (as long as the water is there in the first place). It is still true that the oxygen which can, in principle at least, be readily extracted from lunar rocks is 89% of the weight of water. However, scenarios in which lunar colonies depend on regular supplies of hydrogen from earth don't seem terribly promising.

Partly because of the absence of water, and partly because of the lunar gravity field which makes it a bit hard to get to, attention has been only recently focused on a part of the solar system which may have the advantage of accessibility and of high hydrogen—or water—content. Like the moon, the asteroids were far from the mainstream of astronomy for most of the 20th century. Yet they, or the moon, or both, may be the key to settlement of the inner solar system.

CHAPTER 10

Near-Earth Asteroids

Gold Rush Country?

The exploitation, as well as the exploration, of the near-earth asteroids has emerged only recently in discussions about the future of the space program. Brian O'Leary, John Lewis, and William Hartmann, among other scientists, have called attention to this rather unlikely group of solar system objects as possible destinations and sources of extraterrestrial resources. We know relatively little about these objects, so that much of what is written about the asteroids is somewhat speculative. The possibilities, though, seem very intriguing.

The asteroids, or minor planets, are a group of thousands of small objects orbiting the sun in the same way that the planets do. Ceres is the largest and the first one to be discovered, and it's only about 600 miles in diameter, one quarter the diameter of the earth's moon. Most are far smaller. The vast majority of the asteroids orbit between Mars and Jupiter, but about a thousand have been bumped into the inner solar system and follow orbits which are comparable in size to the earth's. It is these near earth asteroids which have recently become a major focus of interest, because they are so easy to get to and because they may be much better space resources than the moon is.

The near-earth asteroids have a lot going for them as potential sites for space exploitation. The two major disadvantages of the moon as a resource base (the absence of water and the presence of a strong gravity field) don't apply to these asteroids. Most of them are very tiny—only a few miles across—so it takes very little rocket fuel to lift material off of the asteroid's surface. There are indications that some of them, at least, may contain considerable quantities of water.

Asteroids were once quite infamous. Astronomers called them the "vermin of the skies." One reason for such disdain was that an astronomer working on something else which he thought was more interesting found them an unwelcome diversion. A scientist who came across one of these objects on a photograph taken for other purposes was driven by a scientific conscience to spend hours measuring the position of this uninteresting, tiny rock, so that someone else could compute an orbit for it to be sure it could be found again. Even the discoverer's privilege of naming an asteroid was insufficient incentive to attract many astronomers to studying asteroids. In the 1970s, the author of an article classifying astronomy textbooks regarded a book as being old-fashioned if it devoted much space to the asteroids. Sometimes it seemed as though they had vanished from the solar system, disappearing from scientific sight.

But now, what a change! These Lilliputian rocks and iceballs, once in an astronomical backwater, are now at center stage. Science, like women's clothing and hair styles, has its fashions too. The asteroids have become *haute couture,* just as the moon did in the Apollo days. People who study them can get grants, hold conferences, and even propose space missions which have a chance of being paid for, built, and eventually flown. An asteroid mission is next in line in NASA's planetary program, though thanks to *Challenger* it has been waiting for a definite go-ahead for several years.

INTRODUCING THE ASTEROIDS

Asteroids, or minor planets, have long been regarded as an unimportant constituent of the solar system. Astronomers share a fascination with bigness. All but the largest two known asteroids could fit inside the state of Texas. Either their tininess or their abundance discouraged interest.

The upsurge of interest in the asteroids has both scientific and economic roots. The scientific impetus to this revival has been the recognition that many asteroids have surfaces which have remained unaltered since the beginning of the solar system. The study of asteroids, then, offers insights about earth's origins rather than about how geology operates in different settings. Perhaps more important to this book has been the "economic" basis, where the economics are very futuristic. The possibility that asteroids may be one of the most useful and accessible sources of extraterrestrial materials was first articulated in a 1976 summer study about space colonies.[1] Much remains to be done.

Orbits

Most asteroids orbit in the asteroid belt between Mars and Jupiter. The pattern of planetary orbits suggested to a number of 18th century astronomers that a planet of some sort should orbit between Mars and Jupiter. In Palermo, Sicily, astronomer Giuseppe Piazzi was foregoing some New Year celebrations on the evening of January 1, 1801, and discovered a moving, tiny dot. Its orbital speed just matched the speed of the hypothetical missing planet. This tiny dot did indeed "fill the gap" between Mars and Jupiter, but its small size—it is now known to be 615 miles (1025 km) in diameter— made it quite different from the known planets. The smallest of the planets known in 1801, Mercury, has a diameter of 2880 miles (4800 km), nearly five times the size of this asteroid. Piazzi named the asteroid Ceres, and it is now known as "1 Ceres" where the number indicates the order of discovery.[2]

Asteroids orbiting between Mars and Jupiter might be of considerable scientific interest, but they are so far away as to be of little practical use. A new, important type of asteroid was first discovered in 1873. The first member of the group of "near-earth asteroids," called 132 Aethra, has an orbit which crosses the orbit of Mars, bringing it outside the asteroid belt into the inner solar system. Some asteroids penetrate even deeper toward the sun. 1862 Apollo (discovered in 1932) has an orbit which crosses the earth's orbit, 1566 Icarus crosses Mercury's orbit, and 2062 Aten is, on the average, closer to the sun than the earth is. The similarity between these asteroids' orbits and the orbits of earth and Mars makes them much easier to get to than the asteroids in the belt. While asteroid aficionados subdivide this class of objects into Amors (Mars-crossers), Apollos (earth-crossers), and Atens (near-earth objects which don't go beyond Mars's orbit), the name for the group overall, which I'll stick with, is the near-earth asteroids.

Size and Composition

A few asteroids are quite large; 30 bodies are more than 200 km in diameter. The vast majority are very tiny, and if we were able to see more of them we would, no doubt, discover hundreds of thousands of kilometer-sized rocks. It's easiest to discover the small ones if they are near-earth asteroids, and since 1973 astronomers Eleanor Helin and Eugene Shoemaker have conducted a small-scale search for these objects. Shoemaker and Helin have increased the number of identified near-earth asteroids from a handful to nearly 100; they estimate that there are about a thousand which are more than 0.6 mile (1 km) across.

Although the asteroids are quite varied in composition, only a few of the broadest classifications are of interest from the viewpoint of space resources. Some are big lumps of iron. Some resemble terrestrial igneous rocks, rocks made when melted rock cools. Most fall into neither of these categories, being very dark in color. These dark asteroids resemble a lump of coal, reflecting all colors of light equally badly. Technically, these are called "C-type" asteroids with various subcategories (CI, CO, CM, CV); I'll informally refer to them as the dark ones. What are these very dark asteroids made of?

Meteorites

Fortunately, nature has provided a way for us to examine the asteroid belt directly. A rock traveling through interplanetary space can run into the earth. If it is big enough, it will survive its fiery passage through the earth's atmosphere and reach the ground. Such rocks, called "meteorites," are the only samples of the cosmos from beyond the moon which have made their way to terrestrial laboratories.

One class of meteorite, the *carbonaceous chondrites*, have colors that resemble the dark asteroids. These meteorites have attracted considerable scientific attention because their chemical composition resembles the composition of the sun, without the helium and hydrogen which are the dominant elements in the sun's makeup. This resemblance caused scientists to believe that these objects were very primitive, and could provide insights into how the solar system formed.

Now these objects seem even more exciting because of the possible connection with the very dark asteroids. The similarity in spectrum between the carbonaceous chondrite meteorites, and the likely origin of meteorites in the asteroid belt, suggest quite strongly that the dark asteroids are similar in composition to the carbonaceous chondrites. A space mission could certify this similarity, or discover that the dark asteroids are extinct comet nuclei which have accumulated a dark crust around them; the close-up photos of Halley's Comet sent back by the European and Soviet space probes show that this comet nucleus is covered by a dark crust of unknown composition.

If—and this is a pretty big "if"—the dark asteroids are like the carbonaceous chondrite meteorites, they are quite interesting objects indeed. They contain primarily clay and about 2–5% carbon.[3] Many of the clay minerals include water of hydration, that is, water chemically bound to the mineral. The amount of water can be as high as 10%. Space settlement enthusiasts, as thirsty for water as parched desert travelers, positively drool, visualizing these dark asteroids as potential oases in outer space.

Another class of asteroids, the metallic ones, might also have some economic value. The most easily recognizable meteorites are the "irons," rocks consisting of a metallic iron-nickel mixture. Some of the larger asteroids (specifically 3 Juno, 6 Hebe, 7 Iris, 8 Flora, and others) reflect light in the same way that the iron meteorites do. If these asteroids are indeed mostly metallic, they could be used to build spacecraft in space, avoiding the need to haul metal from the earth's surface.

Motivated, perhaps, by the accessibility of the near-earth asteroids, scientists are beginning to study their composition. Lucy-Ann McFadden's 1983 Ph.D. dissertation at the University of Hawaii was the first attempt to answer this question. She measured the colors of the brightest dozen and a half asteroids, and found examples of all of the major meteorite types except for the iron ones. Based on her work, planetary scientist William Hartmann estimates that about 60% of near-earth objects, including burnt-out comets, are in the dark class, having carbonaceous chondrite composition, which suggests an abundance of water.[4]

How did the asteroids get this way? Asteroid specialists disagree about details, but a basic picture has evolved over the last two decades and seems reasonably secure. The very existence of iron meteorites demonstrates that they must have, at some time, been in the core of a rocky object that was big enough to be molten from center to surface. Such an object, called a "parent body," might well be rather similar to some of the larger asteroids which still remain. A violent collision between this parent body and another one would smash the two objects to bits, exposing fragments of the iron/nickel core, which would eventually fall to earth as iron meteorites.[5]

ACCESSIBILITY OF THE ASTEROIDS

One of the attractive features of the near-earth asteroids is their small size. While it makes them hard to discover, it makes them easy to take off from. The velocity change required for a round trip to many near-earth asteroids is less than the velocity needed to get to the surface of the moon. Most of the delta-vee needed to get to and from an asteroid is consumed by the need to match velocities with the asteroid on the outbound trip; typically, this delta-vee is around 5 km/sec. Coming back is spectacularly easy, since you can use aerobraking to slow you down at earth approach.

The round-trip delta-vee figures mentioned earlier of 5.5 km/sec for a typical near-earth asteroid and 4.5 km/sec for the best case, 1982 DB,[6] indicate how easy the asteroids are to get to. Comparable round-trip values are 7.4 km/sec (Phobos), 8.7 km/sec (lunar surface), and 10.5 km/sec (Mars). If appreciable quantities of mass are to be extracted from these places, the asteroids look even better, for in this case it is the return trip delta-vee which matters. Sending a virtually empty spacecraft to some solar system destination to pick up some stuff and return laden with cargo can produce a mission profile in which most of the rocket fuel is used up on the return trip. The return trip from near-earth asteroids requires a minimal delta-vee of 0.5 km/sec, compared with 1.8 km/sec (Phobos), 3.1 km/sec (lunar surface), and 4.8 km/sec (Mars). If rocket fuel were the only thing that matters, the asteroids would be clear winners hands down, in spite of the fact that they are tens or hundreds of millions of miles away and it's only a quarter of a million miles to the moon.

Two other considerations could make the asteroids less attractive. The journey lasts a long time, the better part of a year, and the return trip is equally long. Journeys to the moon take less than a week. However, in an era when robot spacecraft will undoubtedly do the first exploratory missions and may well do the mining too, trip time doesn't matter all that much. Even if we're talking about sending people to asteroids, trip time still is likely to be unimportant. If the whole space settlement scenario makes sense at all, we will have learned how to solve the problems of long-duration spaceflight and closing life support systems, so that a long-duration flight need not take any more resources than a short hop.

A second problem with the asteroids as locales for space resource exploitation is potentially more serious. Achieving such low delta-vee's requires a traveler to begin a journey during particular times or launch windows. It is only at certain times that the origin and destination points are aligned so that a minimal delta-vee can be achieved, so that a spacecraft can follow a gentle loop, the Hohmann transfer orbit, which arrives at the destination at the same time as the target asteroid is there. Launching a space mission outside a launch window imposes a very substantial penalty, as discussed in Chapter 7. Traveling to the moon from low earth orbit (LEO) is quite easy; an appropriate launch window can occur every

90 minutes, each time that the spacecraft goes around the earth. From the time perspective we're talking about, the launch window to the moon is always open.

For the asteroids, however, the launch windows are far narrower and less frequent. The frequency of opportunities depends on the particular asteroid. Former astronaut Brian O'Leary argues that opportunities for the exceptionally favorable target 1982 DB recur at decade intervals.[7] Launch windows for other asteroids are 2–3 years apart, as is the case for Mars. A mining mission or, more critically, a resupply mission to an inhabited base on an asteroid must be made ready and launched within a month or so of the center of the launch window, or else the required expenditure in rocket fuel goes up enormously.

Whether the perpetually open lunar launch window gives the moon a big advantage is an open question. From the contemporary perspective of the slowly reviving space shuttle program, where postponements and delays of weeks or months occur regularly, it seems almost impossible to target a time interval of a month and expect to get a particular mission flying in that time. However, this may not always be the case.

The potential advantage of the asteroids compared to large bodies like the moon or Mars could be even greater if some of the advanced propulsion technologies discussed in Chapter 8 come to pass. Electrical propulsion or solar sailing only works in interplanetary space; it's likely that chemical rockets will remain the best way to escape from a strong gravitational field for a long time to come. Much of the energy required to bring materials back to low earth orbit from the lunar surface is used to escape the moon's gravity field, and solar sails or ion engines cannot be used for this purpose. These advanced propulsion technologies could be used for virtually all stages of a round trip to an asteroid, while they could not be used to land on or take off from the moon.

ASSAYING THE ASTEROIDS

Decades ago, a gold miner who found a vein in the Klondike took his nuggets to a local chemist to have them assayed, to have their value determined. We can try to make a similar assay of the asteroids based on their connection with meteorites and our fragmentary knowledge of their properties. What follows is an inventory of the resources which scientists expect to find among the near-earth asteroids.

Water

Whether the dark asteroids resemble carbonaceous chondrite meteorites or whether they are extinct comets, water may be quite abundant on them. If the dark asteriods are indeed analogous to the carbonaceous chondrite meteorites, the 1–10% water of hydration can be driven off simply by heating the asteroidal material to a modest temperature of 250–300°C. The water comes off as water vapor; making liquid is simply a matter of collecting the vapor and then cooling it.

If the dark asteroids are ex-comets, water is likely to be even more abundant and easily extractable. The images of Halley's Comet, and analysis of what happens to extinct comets, suggest that these objects are mostly icy (or watery) nuclei covered with a relatively thin blackish crust of carbonaceous material. Extraction of water from such an object requires you to drill a hole through the crust. It might not even be necessary to heat the interior if the temperature and pressure conditions on the object were such that the outer layer of the core is liquid. Water could be extracted until the entire object has been consumed; a half-mile cometary nucleus could keep space settlements going for centuries.

Once this water is brought to a space settlement, humans can use it in many ways. Clearly space settlers can drink it, wash with it, and irrigate plants with it. No matter how hard we work on closed life-support systems, some water will be needed to keep them going. Water from asteroidal materials could be used as a source of rocket fuel. The most economical rocket fuel is hydrogen and oxygen. An apparatus which passes an electrical current through water can easily dissociate the water into its component elements, making hydrogen and oxygen. Solar panels can be used to provide the electrical current. With this relatively simple scheme, dark asteroids could become the refueling stations of the future, and there would be no need to lift rocket fuel from the surface of the earth at great expense.

Hydrogen and oxygen might be used for other purposes, too. If for some reason lunar bases and asteroid mines existed side by side in the 21st century, asteroids could provide the hydrogen necessary to extract oxygen from the lunar rocks. These gases might also be used in mining processes on the asteroid itself.

Iron and Nickel

The asteroids contain pure, unoxidized iron and nickel, suggesting another comparatively simple scenario for the exploitation of the asteroids. On the metallic asteroids, mining iron and nickel is as simple as using a robot lander to pick up rocks. Other asteroids contain these metals in the form of tiny grains, and iron extraction is still easy because iron is magnetic. A robot prospector can simply crumble the soil and extract the metallic grains with an electromagnet. Move the magnet near a collecting device, turn it off, and turn a magnet at the bottom of the collector on. Because asteroids have negligible gravity, simply holding the magnet above the collecting can and turning it off won't work. In any of these cases, the robot prospector can then bring the iron and nickel back to a space settlement in near-earth orbit.

Iron and nickel are very useful metals, and have been for a long part of human history. At the dawn of civilization, humans first made tools of stone, then of iron. We now use iron mostly in the form of steel, where iron is alloyed with carbon and with other metals like nickel, molybdenum, and chromium. While it might be difficult to produce stainless steel, which contains 10–30% chromium, in space, there is no oxygen in space so you don't have to worry about rust. An alloy of iron, nickel, and carbon would be quite easy to produce from asteroidal materials.

While spacecraft made of aluminum are definitely lighter than spacecraft made from iron and steel, it might well make sense to use asteroidal materials to construct some structures in space. Space stations and space colonies, in particular, stay put in an orbit once they are fabricated. The weight penalty resulting from the use of asteroidal iron and steel wouldn't make much difference in this case.

Nitrogen and Carbon

Nitrogen and carbon are two resources which are useful for living in space, though less essential than water. The lunar soil contains very little nitrogen and carbon, because both tend to form compounds which are gaseous and which escape from the lunar surface. However, these elements are likely to be abundant in the near-earth asteroids. The carbonaceous chondrites are 3% carbon and 0.1% nitrogen. Again, I'm assuming that the dark asteroids resemble the carbonaceous chondrites. If they are extinct comets, we can't produce an assay in the form of percentages, but cometary gases contain lots of carbon and nitrogen and a reasonable guess is that their nuclei contain these elements too.

What would the nitrogen be used for? The disastrous Apollo fire in 1967, where a tiny short circuit turned the Apollo cabin into an inferno in the pure oxygen atmosphere, demonstrated that pure oxygen atmospheres are quite dangerous. While the Apollo spacecraft continued to use pure oxygen, the space shuttle and the Soviet spacecraft use a mixed nitrogen/oxygen atmosphere. Because no spacecraft will be completely airtight, nitrogen and oxygen will be needed in order to top up the air supply. Oxygen is easily obtained by electrolysis of water from wherever the water is obtained. Nitrogen must come from a separate source, and the asteroids could easily provide it. Of course, oxygen and nitrogen can also be obtained from the earth's surface, but that's what we're trying to avoid.

The carbon found in asteroids could find several uses. The most likely chemical process for carbon extraction would produce it in the form of methane and carbon monoxide. Carbon monoxide is an essential ingredient in one proposed metal mining scheme, described below. Methane could be used as a rocket fuel. While methane does not, pound for pound, produce as much thrust as hydrogen does, its ease of handling and its accessibility may give it some distinct advantages in the space economy of the 21st century. It might make sense to polymerize asteroidal methane into oil for use as a lubricant, or polymerize it even more to make plastics. It could also be used in steel-making, as described earlier.

If future space settlements produce most of their own food, the settlers could use carbon and nitrogen. Little serious thought and no experimentation have addressed the question of growing food in space; those plants which have been grown have been seen primarily as sources of oxygen, not edibles. A living human gets rid of these elements by breathing out carbon dioxide, by urinating, and by excreting. I find it hard to believe that even an advanced closed life-support system could recover all of the carbon and nitrogen from these sources, and suspect that an additional source of these elements would be required in order to support a space settlement.

Platinum and Gold: The Klondike of the Skies

The asteroidal resources discussed so far are all fairly common on earth. The primary purpose of obtaining them from the asteroids is so that future space settlements won't have to pay the huge freight bill to lug them up from the earth's surface. Water, nitrogen, carbon, iron, and nickel are all common enough on the earth's surface that it wouldn't make sense to import these substances from space. Mining the asteroids for these elements would make life in space cheaper, but it wouldn't provide any return to the earth. However, the asteroids do contain something which could provide a very tangible return on investment.

This final class of asteroidal resources is the most speculative, the most important, and the most exciting to me. Gold played an important role in the Western European exploration of the earth, for it was very valuable, universally salable, and portable. My earlier historical discussion stressed the important role which high-value commodities like gold and spices played in making early exploratory expeditions profitable. Similarly, high transportation costs from low earth orbit to the ground suggest that anything made or extracted in space and sold on earth has to be worth a lot. The exact price depends on the rocket and aerobraking technology of the 21st century, and is hard to pin down precisely. However, I suspect that $1000 per pound as the freight charge from low orbit to the surface of the earth is probably a reasonable guess at this point, and any conceivable space product should be worth at least that much to stay in the realm of possibility.

An important group of strategic elements are the platinum-group metals, including gold, sliver, platinum, rhenium, osmium, iridium, and ruthenium. Many of these elements sell on the metals market for prices of several hundred dollars per ounce, close to $10,000 per pound, putting them within the realm of possibility as the end product of space industries. Gold and silver are the most famous of these, and they do have important industrial uses. However, some of the others are even more crucial from a strategic viewpoint; platinum is a useful example.

Platinum is one of the densest materials known; a cubic centimeter of platinum weighs 21 times as much as the same volume of water. It's more valuable than gold ($540 an ounce vs. gold's $444 per ounce in July 1988). It is used extensively as a catalyst, a chemical which can assist the breakdown of other chemicals. A common use of platinum is in the catalytic converters of automobiles. Its role as a catalyst is equally critical in the chemical industry. Platinum is one of a small group of elements which, in the view of the United States Army War College, represent an Achilles heel of the American economy. We currently use about 27,000 kg of platinum per year. American mines and scrap account for about a tenth of this, and virtually all of the remainder is imported from the Soviet Union and from South Africa. Even if we were to commandeer the entire output of the third and fourth largest producers of platinum in the world (Japan and Canada), we would only obtain half as much platinum as we use now. It's certainly conceivable that either South Africa, or the Soviet Union, or in the worst case both countries could interrupt the supply of platinum.[8]

Chemical analyses done on earth show that the metallic grains found in chondritic meteorites contain platinum-group materials in higher concentrations than in terrestrial ores. If these grains are also found in similar asteroids, then the asteroids will contain platinum nuggets too. There's no reason to suspect that the asteroids are different from the meteorites, but at the moment astronomers can only guess.

How can these heavy metals be extracted from asteroids? A relatively simple chemical process can extract the platinum-group metals from asteroidal materials without too much trouble. This process is used to extract nickel from metallic ores, and is easily adapted to extract platinum-group metals. Passing carbon monoxide gas over meteoritic material at a pressure of a few atmospheres and a temperature near 100°C causes the nickel and iron to react to form gaseous compounds, nickel carbonyl [$Ni(CO)_4$], and iron carbonyl [Fe(CO)5]. Higher pressures can extract cobalt and, finally, the platinum group metals.

Asteroid enthusiasts John and Ruth Lewis describe a scenario in which asteroid mining could not only provide space resources but could even pay for itself. A first application of the carbonyl process yields iron and nickel carbonyl, which space workers can use as a feedstock for space industries (e.g., making more spaceships). A second application of the carbonyl process at somewhat higher pressure extracts the cobalt, which is a strategic metal and should be useful for something in space. The real interest, though, is not in the nickel, iron, or cobalt; it's in what's left. The residue of this process is a mixture of mostly platinum-group metals which, they contend, is worth about $20,000 per kilogram ($10,000 per pound) on the present terrestrial metals market.[9] NASA's Solar System Exploration Committee agrees with the Lewises concerning the potential value of the asteroids, pointing out that "a metal asteroid one kilometer in diameter, if such a thing exists, could be worth more than $1 trillion at current market prices."[10]

MAKING ASTEROID MINING A REALITY

Because we know so little, it's not hard to make asteroid mining sound so easy and attractive that it's surprising that the gold rush to the asteroid belt hasn't already started. But realizing some of the dreams set forth in this chapter requires a great deal more work. The planetary exploration programs of NASA and its foreign competitors have paid scant attention to these bodies. If they are to become a significant extraterrestrial resource, the answers to a number of questions must come out the way that we hope that they will.

Finding More Near-Earth Asteroids

The first near-earth asteroid was discovered a century ago, but accidental discoveries of additional ones added only a few more to catalogues. Eleanor Helin and Eugene Shoemaker's systematic search for near-earth asteroids began very recently in 1973. A small telescope on Mount Palomar discovered 64 of these objects by 1984. Shoemaker estimates that about a thousand large objects, with diameters larger than a kilometer, and 400,000 smaller objects (0.1 km–1 km in diameter) exist in earth-crossing orbits.

Thanks to Shoemaker and Helin's pioneering work, cataloging asteroids is one of those few areas of science where we can be reasonably confident that more money will mean more results. To find earth-crossing asteroids, an astronomer simply photographs large regions of the sky and looks for the characteristic streak left by an object moving through the field of view. A larger telescope dedicated to this search could find more asteroids, particularly the interesting dark ones. Additional work on classifying these asteroids could also be done quite easily. Obviously, we have only begun to inventory the potential resource base in the near-earth asteroids.

Space Missions to Asteroids and Comets

The assertion that some of the near-earth asteroids contain exploitable resources is based on the assumption that the dark asteroids are indeed similar to the carbonaceous chondrite meteorites that have made their way to terrestrial laboratories or on their potential resemblance to cometary nuclei and guesses about what cometary nuclei are like. While that assumption is consistent with what we know, none of the near-earth asteroids is more than a tiny dot in a telescopic photograph. We have no direct measurements of their chemical composition. Scientists are still arguing about whether the near-earth asteroids are more closely related to extinct comets or to asteroids in the asteroid belt.

Planetary scientists generally divide solar system objects into three groups: the inner planets (Mercury, Venus, Earth, Mars, and the moon, conveniently ignoring the fact that strictly speaking the moon isn't a "planet"), the outer planets (Jupiter, Saturn, Uranus, Neptune, and Pluto), and the primitive bodies (asteroids and comets). The inner planets have received the most attention, largely because they are easy to get to. While the near-earth asteroids are also easy to get to, the primitive bodies have received the least amount of attention in various space programs. The various missions to Halley's Comet and to Comet Giacobini-Zinner in 1985 and 1986 were the first close-up investigations of any primitive body.

Three missions are currently in NASA's planetary science pipeline. The Magellan mission will produce a high-resolution radar map of our sister planet Venus, several years after the Soviet Venera mission produced a similar map which only covered one third of the planet. The Mars Observer mission will investigate the red planet and clear up some of the mysteries about water on it. The Galileo mission to Jupiter will arrive at and orbit the giant planet in the mid-1990s, some 20 years after NASA assembled a team of scientists to design the spacecraft. None of these missions targets the primitive bodies.

The next mission in NASA's pipeline is supposed to be the Comet Rendezvous/Asteroid Flyby (CRAF). Ever since *Challenger*, this mission has fallen just short of the magical "new start" status in NASA's budget. NASA has spent money on conceptual designs for the spacecraft, and has selected a team to build the instruments and plan the mission, but has not yet begun to build the spacecraft and commit the agency to accomplish the mission. Current plans for this mission have a spacecraft flying by an asteroid first, and then matching velocities with a comet. Instruments, including a penetrator to be implanted in the comet's surface, would follow what happens as the comet comes close to the sun, heats up, becomes brighter, and produces the famous tail. The particular comet and asteroid to be visited depend on the launch date, and the launch date depends on when NASA begins to build the spacecraft. As a result, the mission's destination is unknown. While the major focus of the mission is on comets rather than asteroids, it will obtain pictures of the surface of an asteroid.

A second mission, now called the Near Earth Asteroid Rendezvous (the acronym NEAR is particularly appropriate), is also in NASA's pre-*Challenger* list of "core" missions for planetary science. A spacecraft with cameras, spectrometers, and magnetometers would match velocities with a near-earth asteroid and remain with it for several months, at least. Such a mission could at last answer the question of whether these objects are from the asteroid belt or are dead comets, and could clear up many of the uncertainties regarding the similarity between asteroids and meteorites. In 1982, NASA's plans for this mission only called for a visit to one asteroid, but it strikes me that a mission to several asteroids, either with multiple spacecraft or with one spacecraft making multiple visits, would be more useful, both in terms of planetary science and in terms of inventorying extraterrestrial resources.

At the same time, the Soviet Union and the European Space Agency have announced plans to explore asteroids. In press materials distributed at the celebration of Sputnik's 30th anniversary in 1987, the Soviet space agency listed Project Vesta as being launched in 1994. Vesta, a uniquely bright asteroid, might or might not be the destination. The plan is to fly past several asteroids and land on one of them; the press materials mention the possibility of a sample return. This mission may be undertaken in collaboration with the European Space Agency.[11]

The European Space Agency, a consortium of about a dozen European countries, is more confident still. A 1984 planning document, "Horizon 2000," has this to say about the study of primitive bodies.

> Within planetary exploration, this is an area where Europe could take the lead, following on the Giotto mission [to Halley's comet in 1986]. The return of primordial material from primitive bodies, namely from asteroids and comets, constitutes a major theme in planetary science.[12]

The sample return mission is listed as one of four "cornerstone" missions, three of which are to be launched by the year 2000. This plan was drawn up before *glasnost* touched the Soviet space program; the Soviets are currently anxiously seeking international partners, and it's quite conceivable that a sample return mission would be done jointly by ESA and by the Soviet Union. With all these plans being discussed, NASA's CRAF mission could end up being too little, too late.

Mars has two satellites, Phobos and Deimos, which are quite similar to the dark asteroids. In July 1988, the Soviet Union launched two spacecraft toward these two Martian satellites. Plans (to be discussed in more detail in Chapter 11) call for a laser beam to zap the surface of one or both satellites, and instruments to then analyze the composition of the material which is chipped off of the satellite. While information from the Phobos/Deimos mission is only indirectly related to the asteroids, it will still allow some progress on understanding just how small, dark objects evolve.

Can Asteroid Mining Work?

While the chemical processes of asteroid mining are quite simple, the scientists on NASA's Solar System Exploration Committee have identified a number of technological hurdles that remain before dreams of asteroid mining can be translated into reality. While none of these are necessarily insuperable, they both suggest opportunities for current investigation and serve as a reminder that asteroid mining is a dream for the 21st century, not just around the corner.

Handling materials on an asteroid is not easy, especially when it must be done remotely. The material to be mined is most likely a fine, powdery regolith rather than rocks. Handling this stuff, particularly where there's no gravity to make it sink to the bottom of a container or flow through pipes, represents a new challenge. Separation techniques which exploit the density differences of different rock types won't work on an asteroid, since the usual way of detecting density differences is through the action of gravity.[13]

The chemistry of asteroid mining will be somewhat different from that of terrestrial mining. The concentration of useful stuff is, except for lunar oxygen, rather low in comparison with terrestrial ores. Since there is no air or water on an asteroid, many terrestrial mining processes which use huge volumes of water will have to be modified. Hydrocarbon fuels like natural gas or oil, which are often burned to heat ores, aren't readily available.

These potential hazards may be overcome either with ease or with difficulty. A number of scientists, including NASA's Solar System Exploration Committee, believe that exploratory research on these technologies should begin now, and not in the 21st century. Some of these difficulties apply to both lunar and asteroid mines, and some are alleviated by the presence of gravity on places like the moon and Mars. A future discovery that gravity really helps resource extraction could affect the trade-off between lunar bases and asteroid mining. A bit more information on the feasibility of these technologies, which could come from experiments done in the weightless environment of a space station, could help guide future decisions.

* * *

Ten years ago, a small number of visionaries saw that the near-earth asteroids might be a good source of extraterrestrial resources. In the last several years, many people in the space community have realized that the near-earth asteroids have some real potential. We know they contain water, carbon, nitrogen, iron, and nickel, and they may be the most convenient sources of these materials in the inner solar system. A most intriguing, but more speculative, possibility is that extraction of heavy metals like platinum and gold could become a major space industry.

Many of the dreams articulated in this chapter are based on an astonishingly small amount of information about the near-earth asteroids. Until recently, a mere handful of astronomers spent time studying these diminutive, apparently unimportant members of the solar system. Rendezvous and sample return missions to these objects are certainly a prerequisite to any exploitation of asteroidal resources, though we should not take the approach that the only way to find out more about this potential resource base is to launch spacecraft. Modest investments in ground-based research could pay great dividends as we look to the 21st century.

CHAPTER 11

Mars

The planet Mars has always been exceptionally interesting through much of human history. Its coppery-red color makes it quite distinct from virtually all of the other objects in the sky; only a few bright stars are as visibly red. It is the only planet whose surface is at all visible through terrestrial telescopes. The changing colors and enigmatic markings on its surface fueled speculation about life on the red planet. When two Viking spacecraft landed on the surface of Mars in 1976, it became the only body beyond the earth–moon system to be examined at such close range. Its similarities to and differences from the earth make it an excellent test bench for many ideas in the earth sciences.

By the late 1980s, Mars took on yet an additional role. The Soviet mission to Phobos and Deimos signaled clearly that Soviet attentions and ambitions were returning to the red planet. Soviet announcements and stories in the American media fueled interest in sending people to Mars. Such an expedition is a long-term goal of their space program. It also has become clear that if human beings were to eventually settle on any of the solid bodies in the solar system, Mars is a very attractive possibility, though not the only possible destination. For these and other reasons, Mars remains an important attractor for many people associated with the American space program.

Water, the "liquid of life"[1] and the most important of all extraterrestrial resources, was at one time abundant on the surface of Mars, and may well be present on Mars in a reasonably accessible form. A number of geological features on Mars indicate that there were once catastrophic floods on the Martian surface, and it's very likely that at least some of this water still exists in a frozen, icy layer beneath the sandy surface as well as in the martian polar caps. This water could supply future Martian space colonies.

Mars's two satellites, Phobos and Deimos, have also loomed quite large in our thinking about our future in space. The low gravity of these two tiny martian moons makes them quite easy to get to and return from. Indeed, the round trip to Phobos is one of the easiest to make outside of low earth orbit. We know little about these lumps of rock, but what we do know suggests that they may contain at least some water. Phobos and Deimos may turn out to be the most strategically important places in the solar system.

INTRODUCING MARS

Before the invention of the telescope, Mars was simply a planet like any other. Our nearest neighbor moving away from the sun, it takes nearly two years to go around the sun in a rather eccentric orbit. About every two years, the earth catches up with Mars and orbits between Mars and the sun, being as close to Mars as it ever gets. The distance between earth and Mars at these times varies from 35 million miles at the closest to 60 million miles at the most distant. At these times, Mars appears as a ruddy, yellowish-red glowing dot, rising in the east at sunset and setting in the west at sunrise.

After the telescope was invented in the early 1600s, a number of astronomers used the new gadget to observe the planet Mars. Mars is not that easy to look at through a telescope. Even at its biggest, it is only half as large as the more distant planets Jupiter and Saturn. Very often, the only clearly visible features are the polar caps.

These early telescopic observations allowed astronomers to determine some of the basic facts about Mars. On the surface, it is rather earthlike. Its basic properties are summarized in Table 7. Its radius is about half of earth's, but it has only one tenth of the mass. Gravity on its surface is a a little less than half of earth's gravity. If you weigh 150 pounds on the earth, you'd weigh 58 pounds on Mars, appreciably less but still substantially more than zero.

Mars's gravity is strong enough so that it has hung onto an atmosphere. This atmosphere sometimes contains white clouds, and sometimes produces winds which are strong enough to kick up substantial dust storms. Its year is substantially longer than the earth's year, but it does have seasons since Mars's axis is tilted in much the same way that the earth's is.

Table 7. Mars, Earth, and Their Satellites

Body	Diameter (km)	Mass (g)	Earth Masses	Rotation Period (days)[a]	Revolution Period (days)[a]	Surface Gravity (Earth = 1)
Earth	12,756	6×10^{27}	1.00	1.00	365	1.0
Moon	3,476	7×10^{25}	0.012	27.3	27.3	0.16
Mars	6,788	6×10^{26}	0.11	1.02	687	0.39
Phobos	18 × 22	9×10^{18}	1.5×10^{-9}	0.32	0.32	0.0007
Deimos	12 × 13	2×10^{18}	0.3×10^{-9}	1.26	1.26	0.0003

[a]The rotation period of a planet or satellite is the time it takes to spin on its axis. Its revolution period is the time it takes to orbit around the parent planet in the case of a satellite, and the sun in the case of a planet.

Late in the 19th century, the American astronomer Asaph Hall turned his telescope toward Mars and found that it was accompanied by two tiny satellites. Hall had the right to name these moonlets since he discovered them. He chose the names Phobos ("fear") and Deimos ("terror"), natural enough for companions to a planet named for the God of War. Phobos, the largest of these satellites, circles the planet every 7.5 hours, faster than the planet rotates. Deimos ("terror") is smaller and more distant from Mars.

The same person who weighs 150 pounds on the earth and 58 pounds on Mars would weigh 24 pounds on the earth's moon. It takes a substantial investment in rocket fuel to lift this person, or the equivalent amount of mass, off of each object. However, that same person would weigh 0.1 pounds on Phobos (that's about 1.5 ounces) and 0.045 pound—less than an ounce—on Deimos. Someone walking around on Phobos would have to walk very carefully in order to avoid drifting far away from the planet. Depending on the size of space suit the person was wearing, it might be possible to escape from the gravity of one of these moonlets using muscle power, with no rocket assist.

Even once the basic physical properties of Mars were established and its satellites were named, the mysteries of the red planet continued to fascinate observers. However, fascination does not necessarily mean progress. The discovery of the so-called "canals" on Mars by the Italian astronomers Pietro Angelo Secchi and Giovanni Schiaparelli in the 1870s marked the beginning of a 90-year giant step backward in the study of the planet Mars.[2] The Italian word *canale* can be translated in two ways. It can mean "channel," a natural feature of the landscape. It can also mean "canal," a ditch dug by an intelligent being. Either the attractiveness of the alliterative adaptation or the intrinsic interest in intelligent beings led to the interpretation of these fleeting optical illusions as giant waterworks produced by real, live, intelligent Martians.

The "canals" turned out to be optical illusions. The illustration shows one of the best overall views of Mars produced by the Viking mission, a mosaic of 102 photos. The giant linear feature in the middle of the picture, named *Valles Marineris*, is a huge canyon on Mars produced when the surface pulled apart. The pimples at the left are huge volcanoes, and the white stuff near them is a number of clouds in the atmosphere. There are no canals. Where 19th-century observers saw canals on Mars, there are myriads of dark blotches. Evidently, the astronomers couldn't distinguish the individual blotches, and their eyes and brains interpreted these features as best they could, as lines. As astronomer Carl Sagan quipped, there is no question that the canals of Mars are a product of intelligence, but the intelligence was located at the back end of the telescope and not on Mars.[3]

However, it was some time before the true nature of the Martian canals was recognized, and those few planetary scientists who existed in the early part of this century found themselves drawn into unproductive arguments about whether the canals on Mars were "real" or not. The study of planets, and particularly the study of Mars, became an astronomical backwater. For most of the 20th century, the hot field in astronomy was studying stars and galaxies and not arguing with crackpots about whether there were canals on Mars.

With our ability to send space probes to see it up close, Mars became astronomically respectable again in the 1960s and 1970s. In those days, planetary missions flew frequently, and the two dozen photos from the first flyby missions in the 1960s were quickly followed by the hundreds of photos from Mariner 9, an orbiting mission, in 1971. In the American bicentennial year of 1976, two spacecraft called Viking landed on the planet and sent back pictures.

Both Vikings contained miniaturized chemistry labs designed to look for chemical signatures of life. These experiments, which consumed enormous amounts of money, showed tentative yet inconclusive signs of life-like chemical reactions. Most astronomers now feel that there is no life on Mars or, if it's there, Viking would never have found it.[4] Most of the scientific return from Viking came from its photographs of Mars, Phobos, and Deimos, and other measurements of surface properties.

The next two illustrations show the landscape as seen by both Viking landers. The spacecraft had to be landed in relatively flat areas so they would not plop down on the side of a hill and roll over, and so both

Mosaic of about half of the surface of Mars, with north at the top. The linear feature at the center is a huge canyon; the circular features at the left are three high volcanoes. (NASA)

pictures show relatively flat, geologically bland desert areas. The sand dunes show that wind shapes some features in the Martian landscape. The Viking 2 image shows a small channel in the center, which looks like it was cut by a little rivulet.

Photographs from orbit show some more remarkable features of the landscape, which would be nice to view from the Martian surface. Mars, as it turns out, is just the right size to have some spectacular landscape

The surface of Mars, as seen by Viking 1. Sand dunes and boulders dominate the landscape. (NASA)

The martian landscape from Viking 2. A small channel, cut long ago by running water perhaps, winds through the center of the picture. One leg of the Viking spacecraft rests on top of a rock, tilting the spacecraft a little and producing a tilted horizon. (NASA)

features. It's massive enough so that heat leaking out of its interior can stretch the surface out, making giant, globe-girdling cracks or canyons, and can produce volcanoes like the ones we see here on the earth. The earth is so massive that geological activity erases canyons and volcanoes almost as fast as they are formed. On Mars, though, a canyon which stretches a quarter of the way around the planet persists and is not wiped out by a subsequent episode of continental drift. A volcanic mountain which is twice the height of Mount Everest is not worn down by erosion or stretched out to form an island chain. The scenery, when some human looks at it, will be truly spectacular.

GETTING THERE FROM HERE

The Nearness of Mars

Measured in miles, the trip to Mars would seem impossibly long. The easiest path to Mars takes a spacecraft on a long, looping trajectory, curling around earth's orbit, and covering well over a hundred million miles. The trip usually lasts for nine months, though for some launch opportunities it can be made faster. (The Soviet Union's Phobos mission was launched at a particularly favorable launch opportunity, in July 1988, and the trip was to take seven months.) The comparison of this distance with the shorter (quarter-million-mile) distance to the moon creates the impression that getting to Mars is much harder than getting to the moon.

In some ways, it is indeed much harder to get to Mars. A nine-month trip in space is hard. If there are people on board, the spacecraft has to be big enough for them to live in for nine months without killing each other, and it also has to either take along masses of consumables or contain some kind of closed life-support system. We either have to find a way around the bone-thinning problem or construct a spacecraft which has artificial gravity. None of these difficulties is insuperable.

Mars is quite close, though, in terms of delta-vee, the measure of how much rocket fuel it takes to place a pound of payload on the Martian surface. When you add up all of the velocity changes needed to get from low earth orbit to the surface of Mars, it comes to a total of 4.8 km/sec. The comparable figure for the surface of the moon is 5.6 km/sec (as long as the astronaut uses only the minimal amount of fuel to land). In this sense, the surface of Mars is closer than the surface of the moon!

The reason for the funny "nearness" of Mars is that Mars has an atmosphere which can be used to slow a spacecraft down for a Martian landing. When the Apollo astronauts landed on the moon, they had to use retrorockets to slow them down, gingerly working their way down to the lunar surface. Each blast of retrorocket fire used precious fuel, which had to be lugged up from earth at great expense. On Mars, parachutes combined with the same aerobraking techniques which slow the space shuttle down can do the job. Parachutes require an atmosphere to work and wouldn't work on the moon.

When a return trip is included, Mars does indeed turn out to be just a little harder to get to than the moon is. The delta-vee needed to return from Mars to low earth orbit is 5.7 km/sec, considerably more than the moon's 3.1 km/sec. Climbing out of Mars's stronger gravity field takes more fuel. Take the simple approach of adding up all of the delta-vees, and it turns out that going to Mars and back is 10.5 km/sec, while the comparable figure for the moon is 8.6. Mars is farther away, but not by much.

Launch Windows

There is, however, a difficulty. In order to get to Mars easily, with a minimum delta-vee, the spacecraft has to be launched at the right time, when the earth and Mars are aligned in a favorable configuration. There's a short launch window of roughly one month when the required launch energy, or required delta-vee, is pretty close to the minimum. If the mission isn't ready, it's a two-year wait until the next opportunity. Launch windows for the 1990s occur in September 1990, October 1992, November 1994, December 1996, and January 1999.

Similar launch windows apply to return trips. If you want to use a minimum amount of fuel, for a minimum delta-vee, again you have to wait for the right launch window. As it turns out, the timing is far from ideal; return launch windows for the 1990s occur in December 1990, January 1993, February 1995, March 1997, and April 1999. The timing means that any particular mission which goes to Mars and returns can take a very long time. If, for example, a spacecraft were to be launched in December 1996, it would arrive on Mars in August 1997. The next return launch window isn't until April 1999, a delay of some 20 months. Delays for other launch opportunities vary but are still quite substantial. The delays can be reduced quite significantly if a minimum-energy orbit is not used, and it is quite likely that the first trips to Mars with people on board will not require 20-month stays on the Martian surface (see Chapter 17 for some examples of particular mission profiles). But when we consider the long-term prospects for settlement on Mars or on its satellites, the minimum-energy orbits with their associated narrow launch windows are the important ones.

WATER ON MARS

Looking to space travel in the 21st century, one discovery from our automated exploration of Mars stands out above all the rest in importance. Among the tends of thousands of photographs returned from the Mariner 9 mission in 1971 and the Viking mission in 1976 were several hundred showing features on the Martian landscape which could only have been formed by running water. Yet rivers could not run on the Mars of today; the atmosphere is far too thin, cold, and dry. If scientists are interpreting these features correctly, there must have been an early, warm period in the martian climate when rivers flowed. Where was the water then? Most importantly, where is the water now? If it is still on Mars, then maybe we could tap it, use it as an important extraterrestrial resource.

Ancient River Valleys

The illustration shows one example of a landform which could only be formed by running water. Water ran northward, upward in the picture, and the streamflow encountered an obstacle in the form of an impact crater with a raised rim, at the bottom center of the picture. Some nameless chunk of rock blasted this crater out billions of years ago. The crater was a 5-mile wide obstacle in the middle of a huge, shallow stream, now named Ares Vallis, and it produced the form of the remnants of a teardrop-shaped island vaguely similar to the islands which form in the middle of a river. A close look at the photograph reveals many other such islands.

How are these islands made? The streamflow bumps into an obstacle and splits in two, going around each side of the crater. Downstream from the crater, these streamflows don't converge immediately, but leave a region of less rapid flow downstream from the crater. Sand, dirt, and other junk are more readily deposited onto the streambed when the stream doesn't flow as fast. The streamlined, characteristic teardrop shape can only be produced by a fluid which flows freely. Magma—melted rock—just won't do; it's too gooey and sticky. If you look closely at the picture, you'll also see some flow lines which outline the upward, northerly, downhill flow of water.

Another characteristic of water flow, at least on earth, is the existence of tributary systems. Thousands of little streams which begin high

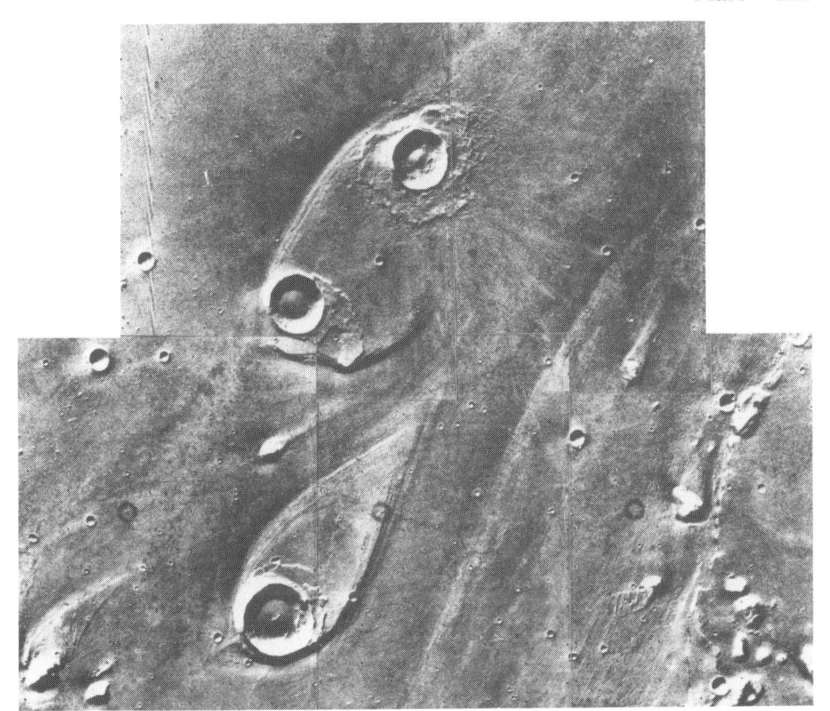

This Viking orbiter picture shows streamflow features in the Chryse basin, near the Viking 1 landing site. Water flowed toward the upper right, and the teardrop-shaped islands are characteristic of features sculptured by water flow. (NASA)

on a plateau flow downhill. Each time two streams meet, they converge to form a bigger stream. The tiny brooks (which, in some parts of the country, are called "creeks" or "runs") merge to form larger brooks and then eventually rivers.

Close analysis of the tributary patterns on Mars indicates, to the trained geologist, that the water in these instances did not come from rainfall. A rain-fed river tends to develop a much more widespread, fan-shaped, complex tributary structure, especially if it traverses a rather flat area. The Mississippi River basin, where water from Pittsburgh, Pennslyvania, Minneapolis, Minnesota, and Butte, Montana, all converges to form the main line of the Mississippi River and flow out into the ocean in Louisiana, is an example of such a river basin. There must be thousands of tiny little streams which all converge into the main stream. On Mars, the ancient river valleys have only dozens of tributaries, not thousands. Why?

The current interpretation asserts that these valleys formed by a process called "sapping," which occurs very occasionally on the earth and is not really well understood. In sapping processes, water buried in layers beneath the surface seeps out in the same way that maple syrup seeps out of trees. The seeping water erodes the water-bearing layer, making it easier for water buried deeper beneath the surface to emerge. Southeastern Utah is one of the few places on earth where sapping occurs. Elsewhere, there's either too much rain or too little groundwater.[5]

The illustration shows a valley on Mars, Ravi Vallis, which was formed by the sudden release of fluid. Streamflow is toward the left in the picture; the crater at the center is about 15 miles across. Water was released from the chaotic ground at the center right of the picture and flowed downhill, creating the streamlines at the center left.

The next illustration shows another type of flow pattern, which again is analogous to similar features seen in a very limited part of the earth. This feature, Mangala Vallis, is one of the most interesting channels on Mars, for a variety of processes have sculpted its valley. The streamlines in the top of the picture and the tributary systems shown in the bottom left are clear evidence for flowing water.

We see similar flow patterns in the channeled scablands in the eastern part of the state of Washington. The remarkable and persistent geologist J Harlen Bretz proposed in the 1920s that the scablands were formed when two ancient lakes (now called Lakes Missoula and Bonneville) suddenly spilled over the melting glacial dam which held its waters in check. Giant flash floods spilled over hundreds of miles of territory and created a remarkably confusing landscape. Bretz's ideas were dismissed for decades. He was finally vindicated in the 1950s and 1960s by additional evidence, most particularly the discovery of Lake Missoula and the recognition that the Great Salt Lake is the parched remnant of a much bigger lake which existed 13,000 years ago.[6]

Are we sure that water is the culprit? The traditional welcome for a new scientific finding is a rather unusual one—a flood of articles criticizing the discovery and proposing alternatives. The idea that water cut the channels on Mars was greeted with interest, respect, and the investigation of many different ideas. Wind, flowing mud, lava, and glaciers have all been reasonably seriously considered as different ways of making the channels. While each of these ideas can explain some features of the Martian channels, only water can account for the streamlines and the teardrop shapes and the tributary structure and the relation between streams and chaotic terrain and all the other little details which add up

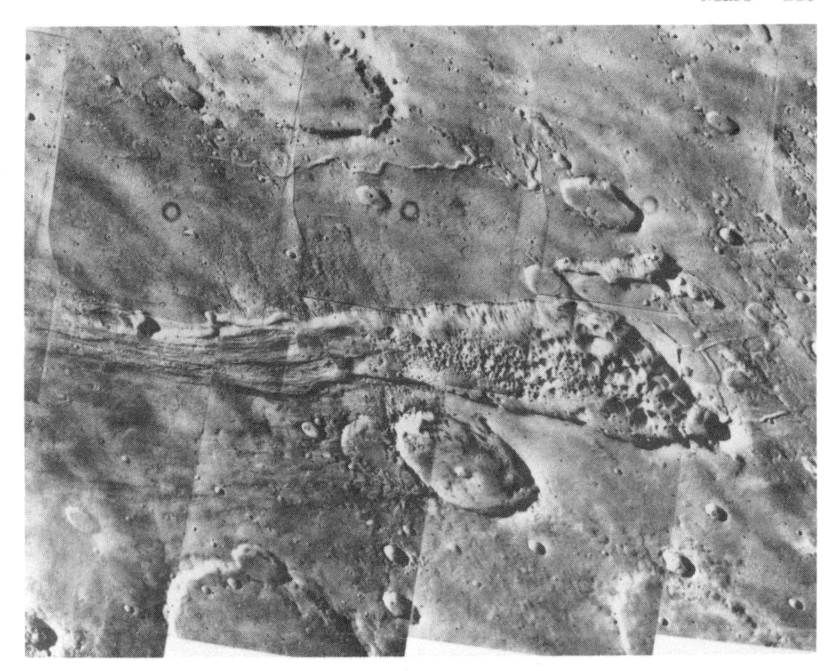

This photograph shows Ravi Vallis, an example of water flow on Mars. Water was released from the chaotic ground at the right, and flowed toward the left. (NASA)

to a secure indictment of water as the eroding agent of the Martian channels.

Other Evidence for Water

Close examination of the pictures returned by the Viking and Mariner orbiters has, many years after the missions, revealed another clue to the history of water on Mars. Near the equator, the plains look very much like terrestrial deserts, huge flattish areas which have been modified by water flow and lava eruptions. When planetary scientists look at pictures of northerly or southerly plains, poleward of latitude 30 degrees on Mars, a number of strange features become apparent. The names given to them—"chaotic terrain" is one which I find to be particularly appropriate—reflect the scientists' puzzlement about the origins of these peculiar landscapes.

Planetary scientist Fraser Fanale spotted a key clue to the origins of these peculiar landscapes in 1976. Fanale recognized that masses of frozen

This photo shows the central region of Mangala Vallis. The braided pattern at the top center is clear evidence of water flow. (NASA)

water submerged in the ground can exist in equilibrium with the cold, thin martian atmosphere at latitudes greater than about 40 degrees. Given the uncertainty in Fanale's estimate of the critical latitude, it may not be just coincidence that these weird surface features only seem to exist in places where ground ice can also exist.

Analysis of the landforms found in the martian "temperate" and polar regions, which contrast with the windswept cold deserts found in the martian tropics, supports the contention that ground ice has sculpted the martian landscape at higher latitudes. As with the analysis of the Martian channels, geologists examine the similarity between the Martian features and the landscape in comparable places like Alaska. Still more evidence for the presence of water on Mars comes from the Viking landers and from some rather technical arguments based on the composition of the Martian atmosphere. One of the Viking instruments, designed to look for organic matter in the Martian soil, produced results which suggest a water abundance of several tenths of a percent to several percent. An X-ray based chemical analysis of the Martian soil, which was only sensitive to oxides of elements with atomic numbers greater than 11, provided indirect confirmation. Carbon has an atomic number of 6, and hydrogen has an atomic number of 1. As a result, this instrument could not directly measure the abundance of carbon dioxide and hydrogen dioxide (i.e., water). However, the sum total of all of the chemicals which this instrument did detect falls short of the total sample weight by several percent, and it's likely that water accounts for the missing stuff.[7]

Instruments exist which could sample water on the Martian surface to depths of about 1 meter below the surface, even from orbit. These instruments detect gamma rays which are emitted from various atoms. Such an instrument could have been put on the Viking lander, but had to be left off in order to make way for all of the biology experiments. One of these gamma ray spectrometers will fly on the Mars Observer mission, expected to be launched in the early 1990s.[8]

Why Did Mars Get Freeze-Dried?

Most of the Martian channels are quite old. It's hard to determine absolute ages of geological features on Mars, because normal geological dating requires a geologist to have rock samples which can be used to determine how far radioactive decay has proceeded in a particular rock. All we can do on Mars is to count craters; a young surface will be smooth and an old surface will be pockmarked. Currently, scientists analyzing crater counts believe that all of the channels are about 3.5 to 4 billion years old, formed in the first billion years of Mars's existence.[9]

Cornell's Steven Squyres answers the question of why streams no longer flow on Mars quite simply. He and many other planetary scientists believe that there never was a water cycle on Mars which was like the cycle on the earth. Each of the channels formed as a series of brief, isolated melting events. A particular channel stopped eroding when water ran out at that particular location.[10] Thus, we don't need to seek to find a cause which would turn off any global water cycle.

This apparently simple idea can't explain the complicated tributary systems of some channels. These could not form on Mars today, even if there was a sufficient source of water. When the complex tributary systems were formed, Mars had to be warmer than it is now, with a thicker atmosphere. The Martian atmosphere may have originally contained 100 times as much carbon dioxide as it does now. That carbon dioxide could have combined with surface rocks to form carbonate rocks; it could have also frozen out to form surface reservoirs of dry ice.[11]

Where Has All the Water Gone?

The existence of the Martian channels and the nature of the Martian landscape at latitudes greater than about 30 degrees strongly suggests that at one time Mars contained a substantial amount of water. It's possible to estimate just how much water existed on Mars by analyzing the channels to see how much water is needed to make them, examining the detailed composition of the Martian atmosphere, or calculating how thick a layer of ground ice should be, if it exists. All three ways of estimating the water content of the primitive Mars come up with answers that differ by a bit, but not too much. If one were to spread the water on primitive Mars all over the planet, the evidence shows that it would be at least 160 feet (50 meters) thick. Some lines of evidence indicate that the primitive water layer could be a mile deep. To set a basis for comparison, if you did the same thing for the earth, the layer of water would be about 3 km thick.[12]

Mars does have two polar caps; they are one of the most easily seen features of the planet when someone looks at it through a small telescope. They expand and retreat with the seasons. Temperatures over the northern polar cap are about −70° Fahrenheit (205 K), well above the temperature where frozen carbon dioxide evaporates in the thin Martian atmosphere, and so it is likely to be composed of water ice. The southern cap is considerably colder in summer, and the part of the cap which persists throughout the year is most likely nearly pure carbon dioxide.

The water which we can see in the polar regions is not enough to account for all the water which existed on the primitive Mars. The amount of water in the northern cap is quite small, only a few meters if the water were distributed over the entire planet. Similar amounts of water are present in the ice-bearing ground which is associated with each polar cap. The polar caps of Mars contain far too little water to explain the channels.

It is most likely that the remaining water on Mars is trapped beneath the surface in a "permafrost" layer. This term has been applied rather loosely in the Martian context, although it has a precise terrestrial definition as being rock or soil which remains below the freezing point for two years. On the earth, where water is relatively abundant, permafrost inevitably contains some water. Permanently frozen Martian ground may or may not contain water.

Enthusiasts of space colonization hope that the Martian permafrost can provide the water supply for a Martian outpost or colony. We simply don't know enough about the Martian permafrost layer to assess the water's accessibility. It's possible, for example, that most of the water exists in ice lenses which are tens or hundreds of meters below the Martian surface. Drilling for this stuff would require that massive drilling rigs be shipped to Mars at great expense. It's also possible that the water on Mars is chemically combined with the Martian rocks to make clays. Extracting water from clays where the water concentration is only a few percent will not be easy.

One can take a much more optimistic view, however. If the water trapped in the Martian surface is relatively pure, then any liquid water, if it exists, will be about half a mile below the Martian surface at the equator. That sounds deep, but the average depth of American oil wells is over a mile.[13] If this water contains dissolved salts, it will liquefy at temperatures lower than 0 °C and will be liquid at shallower depths on Mars. If there had been a seismometer on the Viking missions, perhaps a shallow water layer could even have been identified. Finding the water on Mars, and finding whether it is uniformly distributed over the planet or whether there are particularly good places to drill for it, will require that we understand considerably more about the history of water on Mars than we do now. As it turns out, the presence of water on Mars is one of the most important scientific questions opened up by the Viking missions as well as one of the key unsolved issues which will govern the future (or the lack of a future) of space colonization.

As a result, a number of missions to Mars are being actively discussed by many major space programs. The most immediate is the Soviet Phobos mission, which has as its principal targets the Martian satellites rather than the planet itself. The American Mars Observer, to be launched in the early 1990s, will measure the distribution of water, carbon dioxide, and other volatile compounds over a seasonal cycle. The Soviet Union has announced a 1994 Mars mission which it is still defining more precisely. The 1994 mission will, most likely, include a lander and devices to determine the composition of surface samples. A rover, which can roam 60 miles away from the lander, is being tested in barren areas in the Soviet Union.[14] Another way to provide some mobility is to launch a balloon on Mars. A team of American, French, and Soviet scientists is investigating a plan in which a balloon with cameras suspended beneath it would heat up and float during the day, and drop to the surface at night.

A sample return mission marks the next major step in our plans to explore Mars without sending people there. NASA's Solar System Exploration Committee in the early 1980s, outlining a set of future solar system missions, identified a sample return from Mars as the highest-priority next major mission in the planetary sciences. Such a mission, whether American, Soviet, or cooperative, is quite likely to occur around the turn of the century if not before.

Why is a sample return so important? Can't we analyze the rocks on Mars, without all of the bother of returning them to earth? Our experience with the lunar rocks indicates that returning samples to earth is enormously more valuable than trying to design instruments which will analyze them on Mars. In particular, a scientist can repeat experiments which have puzzling results and can do additional, unanticipated tests if unexpected results appear. NASA distributed the lunar rocks to hundreds of geologists who analyzed them in fancy, sophisticated laboratories. A similar approach could be taken with the rocks and soil samples from the red planet.

MARS'S SATELLITES PHOBOS AND DEIMOS

There has been a resurgence of interest in Mars's satellites because both planetary scientists and space colonization enthusiasts have realized their importance. The scientific interest in them comes from their nature as primitive, geologically unaltered objects. From the perspective of space colonization, they may turn out to be some of the most important pieces of real estate in the solar system. All this for two rockballs which, prior to the Viking mission, were two tiny dots found near Mars which had names and nothing more.

As can be seen in the three accompanying illustrations, craters are ubiquitous on these small objects. Neither of these moonlets is round. The surface of Deimos looks smoother than that of Phobos, but this contrast is illusory because the Viking orbiters approached Phobos at considerably closer range than they approached Deimos. The surface of Phobos is marked by a number of grooves, which all radiate away from a huge crater called Stickney, a 6-mile crater which is almost as big as Phobos itself. The impact which produced Stickney almost, but not quite, broke the satellite into tiny pieces.

Viking orbiter photos, closely analyzed, show that both Phobos and Deimos absorb all but 5–7% of the light which strikes them. Their colors

The best photograph of Deimos was obtained by Viking Orbiter 1 when it was 2050 miles away. Craters as small as 300 feet across are visible.

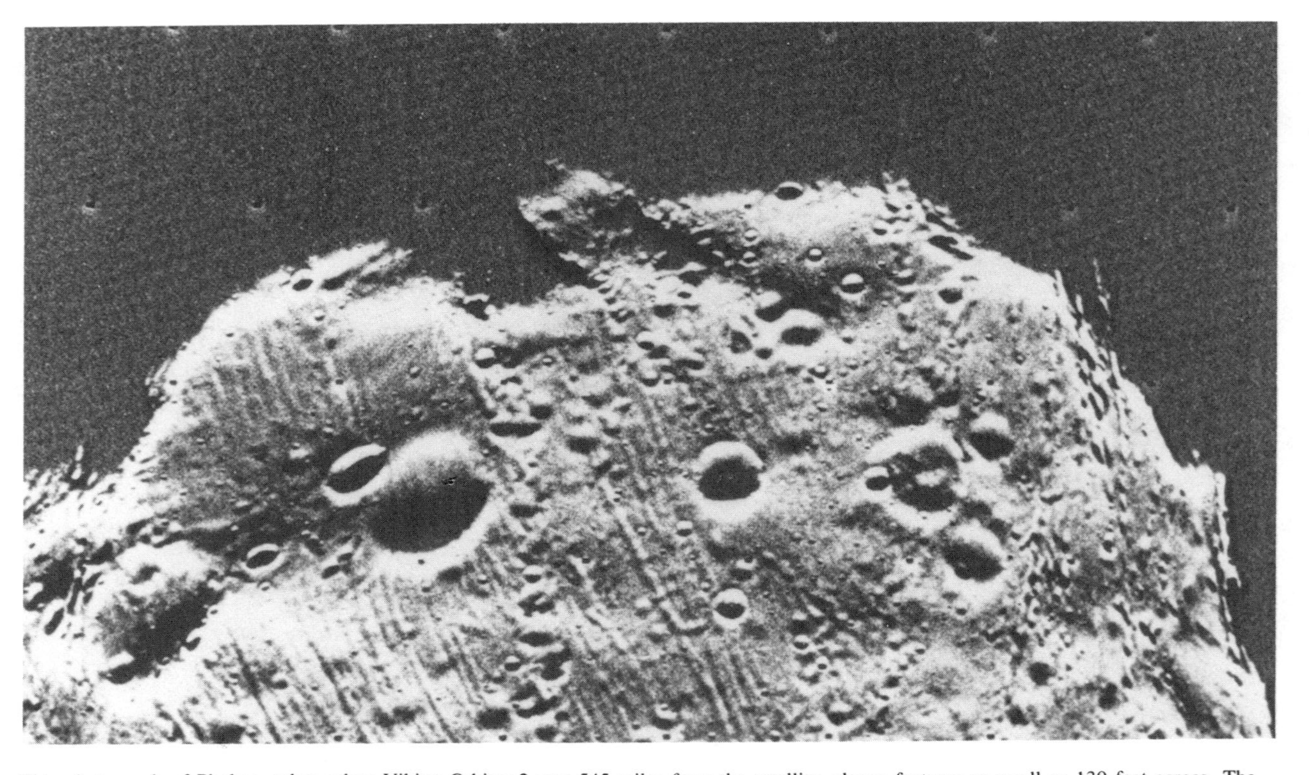

This photograph of Phobos, taken when Viking Orbiter 2 was 545 miles from the satellite, shows features as small as 130 feet across. The striations apparently were formed when an impact produced a huge crater, invisible on the other side of this moonlet. (NASA)

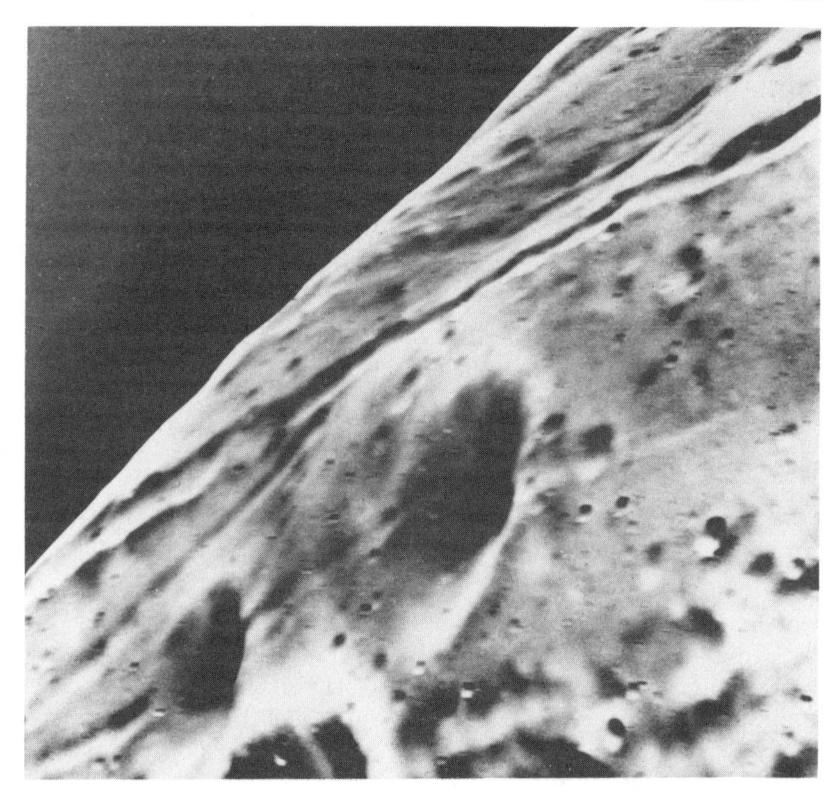

Viking Orbiter 1 captured Phobos close up when the orbiter was only 75 miles from the satellite. The smallest identifiable surface feature is about 50 feet across. (NASA)

are quite neutral, similar to that of the carbonaceous chondrite class of meteorites and the very dark asteriods. Their density is quite low, less than 2 grams per cubic centimeter (g/cm^3). Ice has a density of somewhat less than 1 g/cm^3, while ordinary rock has densities of several grams per cubic centimeter. The low density is evidence that these objects contain a significant amount of volatile material, and water is an obvious candidate. Carbonaceous chondrite meteorites also contain water, organic materials, and possibly other important extraterrestrial resources.

I've just told you practically all that is known about Phobos and Deimos, before the Soviet missions to these satellites in 1989. Why, on the basis of two paragraphs of information, have these tiny, irregular, cratered oversized rocks become the focus of a major Soviet mission?

From a scientific viewpoint, Phobos and Deimos provide a unique perspective on the origin of the solar system. Everyone's interested in origins—how the earth, planets, and solar system formed. Twenty years ago, the antiquity of the moon's surface was touted as an important reason for going there, once the real reason for the Apollo expeditions, pride and power, became submerged. Astronauts searched in vain for "genesis rocks" which would provide us with a sample of what was around when the solar system formed. We learned a lot about how the surface of a modestly massive object like the moon could be altered by impacts and lava flows but we found no genesis rocks. If Phobos and Deimos are giant carbonaceous chondrites, they are primitive; it is possible that the genesis rocks we seek may be found there.

I suspect that the real interest in Phobos and Deimos goes beyond pure science, as is usually the case when a space program changes direction in a major way. The Soviet Union had virtually ignored Mars since its unsuccessful lander attempts in the early 1970s, and then in the late 1980s it sent probes to Phobos and Deimos. These two tiny moons may play a key role in the settlement of the solar system—if we are to settle the solar system. Phobos and Deimos are quite easy to get to, and even easier to get back from, because they orbit around a planet with an atmosphere but they have negligible gravity. If we want to obtain extraterrestrial resources from something in the vicinity of Mars, it makes much more sense to mine Phobos than to mine Mars itself because whatever is extracted from the mine can be sent much more easily to earth. A spring-loaded catapult could get 100 pounds of payload off of the surface of Phobos, while a substantial rocket would be needed to lift the same amount of stuff out of the gravity well of Mars.

The delta-vee figures discussed earlier confirm the nearness of Phobos and Deimos. Because landing on them and taking off from them can be done effortlessly, Phobos and Deimos are nearer to low earth orbit than the surface of the earth's moon, in addition to being nearer than the surface of Mars. The velocity change required to go to Phobos and back to low earth orbit is 7.4 km/sec, less than the 8.7 km/sec for the moon and the 10.2 km/sec for Mars. Their distance from earth is quite comparable to the distances of many of the near-earth asteroids.

In addition, water may be far more accessible on Phobos and Deimos than on the surface of Mars. We know there is water on Mars, but it may be buried deep beneath the Martian surface. Phobos and Deimos may contain lots of water, near the surface, where it could be extracted easily. Phobos and Deimos, then, could become the key oases on the earth–Mars run.

*   *   *

The reddish, dusty deserts of Mars have fascinated astronomers for centuries. More recently, photographs of its surface have shown that some time in the past, the roar of a flash flood could be heard through the thin atmosphere of this planet. While there were, most likely, no living creatures to listen, the water that was in those floods still remains somewhere on the surface of the planet.

Where is that water now? Is it concentrated in the Martian polar regions, close to the surface, easy to extract and use? Or is it buried deep beneath the Martian surface in some inaccessible permafrost layer? Water is the most critical extraterrestrial resource, and so the question of water on Mars is one of the most important questions for us to answer before we know how we can proceed in extraterrestrial exploration.

While Mars is 35 million miles from earth at its closest, the amount of rocket fuel required to put 100 pounds of instruments on the surface of Mars is less than that required to reach the surface of the moon. We have been to the moon already, and so the effort required to reach Mars is well within our grasp. The major barrier is the long time it takes to get to Mars.

For round trips, the Martian satellites Phobos and Deimos are even closer, because little effort is required to take off from them. Phobos and Deimos, furthermore, may be fantastically rich lodes of extraterrestrial resources. Their dark color and low density suggest that they contain a substantial amount of water. They may turn out to be the key oases in the inner solar system.

CHAPTER 12

Where Do We Go From Here?

The late 1980s have been a time in which the future of the space program is a topic of active debate in both of the major spacefaring nations, the United States and the Soviet Union. The spirit of *glasnost* has pervaded the Soviet space program, which has been very open to American participation in a variety of different roles. The Soviets have been very open regarding their short-range plans and have widely discussed the possibility of sending human beings to Mars, possibly in an expedition conducted jointly with the United States.[1] There seems to be a consensus in the Soviet Union that a trip to Mars is the next major step.

No such consensus exists in the United States. Currently, there is an active debate in the United States among the community of space enthusiasts as to whether the next logical step beyond the space station is a base on the moon or on Mars. Some lonely voices suggest that attention be devoted to Phobos, Deimos, and the near-earth asteroids. It is not clear whether top management at NASA has seriously considered these questions, or whether they take seriously the conferences in which the next destination has been actively questioned. The past pattern persists, in that NASA's top management focuses primarily on being sure that the next major megaproject, the space station, is built. However, the size, shape, and timing of the space station will determine whether we are ready to go on to the next destination or whether we have aimed our space station directly down a blind alley.

OVERVIEW OF THE ALTERNATIVES

The phrase "next logical step," often found in NASA planning documents, tends to give a historical inevitability to what is a very clouded situation. Despite the extensive discussion of a trip to Mars, the necessary technology will take decades to develop. I'm convinced that, eventually, we will go to Mars. It is not yet clear whether the Mars trip will be an expedition which will stand alone, both symbolically and technologically, or whether it will be only a part of a broader effort to settle the solar system. Both the timing and the nature of the Mars trip will depend on the mode in which it is conducted. The last several chapters have discussed a number of possible destinations in some detail; each has some advantages and some disadvantages. A summary of these can provide some insight into where we might be headed.

If we are to establish permanent, economically self-supporting habitations in the inner solar system, there is no question that many of the resources we need will have to come from the space environment. I have discussed Mars, the moon, and the asteroids as possible sources for these materials. The Martian satellites Phobos and Deimos share some of the virtues of Mars and the near-earth asteroids. In various conferences, press releases, and other official and semi-official pronunciamentos, various members of the space community have started lining up behind one object or another. The most frequently discussed destinations are the large objects: the moon and Mars. However, Phobos and the near-earth asteroids also have their advantages. In addition, a human expedition to Mars has a symbolic importance, comparable to the moon landing, which can enter the debate in a rather confusing way.

Table 8 summarizes the advantages and disadvantages of particular destinations and sites of exploitable resources. I have listed the available resources in near-earth space to illustrate that there are basically no resources in near-earth space, that self-sustaining settlements in low earth orbit will require water and other necessities of life from somewhere. The table also lists various environmental characteristics, some of which have been discussed in the previous chapters, and some of which are relevant to the possible commercial uses of this particular environment or location.

ACCESSIBILITY

The accessibility of various solar system destinations cannot really be described by one number like distance or even delta-vee, though I feel that delta-vee is by far a better measure of accessibility than miles. We've reached the surface of the moon and returned from it, but in terms of rocket fuel expenditure the lunar surface is one of the most distant of the

Table 8. The Inner Solar System at a Glance

Site	Mars	Moon	Phobos or Deimos	Near-Earth Asteroid	Low Earth Orbit
cessibility					
Round trip delta-Vee from LEO	10.5	8.7	7.4	≈5	Zero
Travel time (one-way)	9 mo	3 days	9 mo	2 yr	10 min to earth surface
Launch windows	2 yr	Continuous	2 yr	2–3 yr[a]	—
ailable resources					
Hydrogen	Yes[b]	NO!	Probable	Likely	No
Oxygen	Yes[b]	Yes	Probable	Likely	No
"Building materials"	Yes	Yes	Yes	Yes	No
Platinum-group metals	Unlikely?	Unlikely?	Unknown	Yes?!	No
vironmental characteristics					
Gravity (relative to earth's)	1/3	1/6	Zero	Zero	Zero
Atmosphere	CO_2	None	None	None	None
Radiation shielding available	Yes	Yes	Yes	Yes	No
Public relations potential	High	Lowest	Middle	Low	Highest

e tabulated figure, 2–3 years, refers to the frequency with which launch windows to one typical near-earth asteroid open .[2] Particularly favorable opportunities to exceptional asteroids may recur more rarely. However, at any time there are me asteroids which are accessible.

hile we know that water exists on the surface of Mars, its accessibility is not known. Permafrost buried hundreds of eters beneath the sands of Mars will be of little use.

various destinations that are being discussed. The surface of Mars is, in this sense, closer, because Mars has an atmosphere which can be used for aerobraking, though on a round-trip basis Mars is a bit farther away than the moon. The Martian satellites Phobos and Deimos are closer than the Martian surface, because the Martian atmosphere can still be used for aerobraking but the surface gravity of these tiny objects is an utterly insignificant barrier to transportation. Closer still are a few of the near-earth asteroids.

Delta-vee alone, however, may not be the whole answer though in my view it is the most important criterion. It makes no sense to even talk about settlement of the inner solar system unless we can develop basically closed life-support systems, but there is still going to be some consumption of water and other necessities of life, and a nine-month trip to Mars or one of its satellites will demand more resources than a three-day trip to the moon. The economy of reaching particular destinations will depend on an imponderable trade-off between 21st-century rocket and life-support system technology.

The moon has an additional advantage in that the launch window is almost always open. Favorable opportunities for reaching Mars or for reaching any particular asteroid recur at intervals of a few years, and completing an economical return trip may also take a few years. If we are to settle or mine beyond the earth–moon system, long trip times and long waiting times will have to be tolerated. It's too early to tell whether these disadvantages will outweigh the ease of reaching the Mars system or the near-earth asteroids.

RESOURCE AVAILABILITY

In resource availability, too, there are some differences which may make various destinations more attractive than others, depending on the needs of the space infrastructure in the 21st century. The lunar rocks contain oxygen and can be used to build bulk shielding for spaceships which will make the Mars run, but unless the polar craters contain readily accessible ice deposits the moon has a distinct liability as a resource base: no hydrogen. Hydrogen is crucial to a future space economy both as one of the constituents of water and as the most economical rocket fuel. We know that some near-earth asteroids, with surface layers similar in composition to the carbonaceous chondrite meteorites, contain water (i.e., both hydrogen and oxygen). Because the surfaces of Phobos and Deimos are superficially similar to the carbonaceous chondrite meteorites, we suspect that they too contain water, but since there are only two Martian satellites (in contrast to about a thousand near-earth asteroids greater than a mile across and uncountable numbers of smaller objects) we can't be absolutely sure that these two objects contain accessible water. There is no question that water exists somewhere on the surface of Mars, but its availability is not assured.

Other resources which might be provided by various destinations include a rather amorphous category of stuff which I call "building materials"—glass bricks for radiation shielding, aluminum and iron for space vehicle construction, and the like. Neither the needs nor the availability of different types of building materials can be forecast at the present time, and so I've contented myself with a simple "yes" entry in the summary table, for all locations except low earth orbit. The importance of that entry is that if humans are to develop a long-lasting presence in low earth orbit, the availability of resources elsewhere and their absence in low earth orbit may provide a powerful motive for extending our presence elsewhere in the solar system, either by establishing inhabited bases or by sending robot miners (or by some combination of both).

The moon, however, is a special case. The likely absence of water on the moon, and its availability elsewhere, may well limit the moon's future value. It may be possible to use lunar soil to make concrete (this has been already demonstrated in terrestrial laboratories) but if the necessary water has to be imported to the place where the concrete is needed, the availability of the lunar soil for cement and aggregate may not be all that valuable (particularly considering that all of that material has to be lugged out of a deep lunar gravity well).

ENVIRONMENT

The environments of the various bodies are also rather different. Mars is unique among the group in that it has an atmosphere. Fifty years ago, this fact encouraged science fiction writers to conjure up dramatic scenes of human beings scampering over the surface of Mars, with little more than an oxygen mask, and sometimes not even that. The immortal novels of Edgar Rice Burroughs (the creator of Tarzan) captured the imagination of a youthful Carl Sagan (and a youthful Harry Shipman too!) by depicting visions of a heroic John Carter rescuing Martian princesses.

We now know better; the atmosphere of Mars is composed purely of unbreathable carbon dioxide, and it's so thin that it provides little protection against the desiccation of human flesh. A space settlement on Mars will need to generate, regenerate, and contain its own atmosphere in the same way that a comparable settlement on the airless moon will. Astronauts on Mars will still need to wear space suits of some sort.

Still, the Martian air is a pretty good shield against cosmic rays. However, bases or colonies in airless environments can provide a similar shield quite easily, simply by burying the base beneath a meter or two of surface material. It doesn't seem to me that the presence or absence of an atmosphere really makes any difference, except that the Martian atmosphere provides a way for incoming spacecraft to slow down via aerobraking without burning rocket fuel. Aerobraking gives Mars, Phobos, and Deimos a slight edge in accessibility.

Mars and the moon also have some gravity, in contrast to the smaller bodies. We do know that the human body has difficulty in adapting to a gravity-free environment. It's not known whether a reduced gravity load on the human skeleton and muscles will still cause the bone thinning which presents such a danger for extended space flight; neither American nor Soviet space station hardware has allowed space medicine experts to experiment with the human response to extended stays in reduced gravity. The short stays of the Apollo astronauts on the lunar surface are of little value in this regard. Gravity may also make mining operations easier, in that fluids will flow down through pipes and terrestrial separation techniques like gold panning can work more easily.

There are a number of other environmental characteristics which will govern what human beings can do in these different places, and it is these which (in my view at least) already suggest some choices of destination. Scientific exploration can take place virtually anywhere, since there is still a great deal to learn about virtually any solar system object. However, a sustained human presence in space is going to require some kind of payoff, some kind of return on the investment in infrastructure. While there are few crystal balls which can forecast the shape of the world economy in the 21st century, the limitations and opportunities afforded by various solar system environments can limit the options rather considerably. The utilization of space is the subject of the next part of this book.

Any major economic or military utilization of space which will require or support the settlement of the inner solar system will, most likely, take advantage of some of the unique qualities of the space environment. An orbiting space station is in free fall around the earth, as is everything in the station. As a result liquid beverages, pencils, nuts, bolts, and other miscellaneous objects don't fall to the ground. Materials research, materials processing, and the fabrication of exotic materials are one widely discussed way to make use of the space environment self-supporting. Other possibilities which have also been discussed are using the good vacuum which space provides at no cost or extracting valuable materials from destinations like the near-earth asteroids.

On these general grounds, the surfaces of large bodies like the moon or Mars look unpromising as likely sites for space industries which will produce earth-bound goods or services. The gravity which might make human life easier in these places makes it harder to develop self-sustaining space industries, which could support the settlement. It may well be that somebody will think of new ideas which would make a settlement which enjoys one third of earth's gravity much more attractive than a settlement with no gravity at all. At the moment, though, environments with zero gravity seem to be significantly more attractive.

PHOBOS, DEIMOS, AND THE NEAR-EARTH ASTEROIDS EMERGE FROM THE PACK

While it is somewhat speculative, these considerations suggest that the smaller, less spectacular, less well known objects in the list of possible settlement sites should be given more attention than the familiar moon and almost equally familiar Mars. On all three general grounds discussed in this chapter—accessibility, resources, and environment—these small objects come out ahead. Unfortunately, our knowledge of these small bodies is pitifully meager; the near-earth asteroids are mere dots on telescopic photographs. Phobos and Deimos are dark, cratered, irregular worlds which have only begun to emerge as a focus of attention in the Martian system. Why are these small bodies so special? Let me count the ways.

First, spacecraft need less rocket fuel to get to them and get back. While trip times are long, trip length is not a problem for robot miners. If the whole space settlement scenario makes sense anyway, we will have to have developed closed life-support systems which can make long trips possible. Small bodies as resource sites have another advantage which the round trip delta-vee figures don't show, namely, that it takes very little rocket fuel to transport materials from them back to the earth. Consider the following scenario: A 10-ton robot (or piloted, if you wish) miner with a huge empty tank travels to Phobos. Phobian ice is melted, extracted, and put into the tank; when full, the miner weighs 100 tons. The delta-vee on the return trip is then the primary determinant of how much rocket fuel you need, and Phobos has a threefold advantage over Mars because practically no rocket fuel is needed to lift off of the surface.

Small bodies have a second advantage, particularly over the moon, in the types of resources which are available. If humans are to settle the inner solar system, water will be as precious a resource as it is to people who live in the desert. "Closed" life-support systems will never be completely closed, and water will be needed to sustain life. If we can't find water on Phobos, Deimos, or the earth-crossing asteroids, the next nearest source of frozen water is one of the outer satellites of Jupiter. We do have evidence for water on Mars, though its availability is uncertain.

The third advantage of the smaller bodies—the greater likelihood of commercialization—is even more uncertain than the second, but it's still there. The lack of gravity, accessibility to the vacuum of space, and the possibility of nuggets of platinum group metals all exist on the small objects and don't exist on the big ones. Skeptics could come up with a counterargument for Mars on this point—perhaps local sources of valuable stuff do exist on Mars, and we don't know about them. Lunar advocates can appeal to the possible use of lunar materials as radiation shielding for Mars craft or as the oxygen component of rocket fuel.

A Space Settlement Scenario

Let's consider all of the optimistic ideas which have been mentioned so far and speculate how they might fit together, picking a date around the middle of the next century as a target for establishing a self-sustaining space industry. For the purposes of this section, I'm assuming that space industrialization does happen and that the cost of living in space does plummet to a minuscule fraction of the present value. Recalling earlier discussions, both of these conditions must apply if space settlement is to make any sense anyway. The question now is, if space settlement occurs, where's the next logical destination beyond low earth orbit?

In this view, a number of fairly substantial space stations will exist in low earth orbit, serving as contract research laboratories to study materials processing in the low-gravity, high vacuum space environment, making high-quality memory chips for use in everyone's desktop supercomputers, and doing biological and medical research in processes such as bone thinning which are accelerated (and thus easier to study) in space. Sales of these services could justify and support building and maintaining the infrastructure to support a settlement in low earth orbit. The low orbit settlements can also make money by fixing communications satellites and other earth-orbiting hardware.

Given the right circumstances, the extension of these settlements beyond the comfortable, relatively radiation-free environment of low earth orbit could occur both because human beings have an exploring spirit and because the settlement needs consumable materials, especially water. The next natural step, then, would be establishing a human-tended base on some water-bearing object in space. This would be justified by the relative ease of bringing water and other needed materials back from this location.

It would thus seem that the natural extension of human activities beyond low-earth orbit would be to those bodies which provide water: Phobos, Deimos, and the near-earth asteroids. Perhaps a variety of remotely operated water mines would make the most sense; limiting one's choice of bases to the two Martian satellites would have the drawback that supplies could only be obtained at two-year intervals when Mars and the earth were aligned properly. The establishment of bases on a number of selected near-earth asteroids in addition to Phobos and Deimos could keep the supply lines open more frequently. Any particular base would only be accessible at two-year intervals, but with a reasonable variety of bases supply ships could arrive at low earth orbit at reasonably frequent intervals, once a month if necessary.

Eventually, the human colonies in low earth orbit would then be supported by a network of automated mining bases, each of which was visited at two-year intervals, scattered among the near-earth asteroids and including the closely related objects Phobos and Deimos. It's quite possible that other uses could be found for these bases. Some asteroids might be tiny bits and pieces left over from the collision of two larger parent bodies, containing chunks of stuff which represents the core of some planet. These objects would be certainly worth studying at close range, since they provide a direct view into a planet's guts. More commercial possibilities include the importation of iron and nickel from the asteroids to low earth orbit for use in building spacecraft. In the most speculative vein, we can dream about mother lodes of platinum-group minerals which could be exported from space back to Mother Earth.

While the previous paragraphs discussed one particular speculative scenario, there is a certain logic to it which may well be preserved even if the future pans out quite differently. Small objects like Phobos, Deimos, and the near-earth asteroids are far more useful than big monsters like Mars and the moon. If cheap, low-thrust spaceflight technologies like ion propulsion and solar sailing come to pass, the attractiveness of these small bodies is even greater.

The Lunar Base

One reasonably clear conclusion comes from the previous speculation. I find it very difficult to see how a lunar base could provide any commercial return unless it turns out that the polar lunar craters contain ice deposits. The advantages of the moon's proximity and frequent launch windows pale before the problem of just what you do with a lunar base once it's constructed. Resources which exist in the lunar rocks require the importation of other materials to become useful. For example, if humans want to use the oxygen in the lunar rocks for water or rocket fuel, hydrogen must be brought to the moon—from where? If the hydrogen comes from a near-earth asteroid, it's just as easy to pick up the oxygen too and forget the lunar base. If hydrogen comes from the earth, the cost will be immense.

While the small bodies seem to have the edge over the surface of Mars, the case is less clear-cut. The surface of Mars does contain the water which spacefaring humans will need. Its atmosphere can help us land people or equipment on Mars with minimal expenditure of rocket fuel. We need more information in order to make intelligent choices.

THE TRIP TO MARS

How does a trip to Mars fit into all these considerations? The Soviet launch of the twin probes to the Martian satellites in 1988 was accompanied by extensive discussion of a possible human expedition to Mars. The human expedition to Mars, which I think will certainly take place before the end of the 21st century if not well before that, will not necessarily establish Mars as the premier space destination for the 21st century. There's no question that the Mars trip is a symbol of human achievement, since Mars is the only planet in the solar system which we can realistically hope to step on. The practical significance of this trip will depend on the shape of the 21st century's space program. Several possibilities suggest themselves:

It could be part of an effort to establish a base or even a colony on the Martian surface. This is the scenario implicitly envisioned by many space enthusiasts, but such a scenario presumes that it makes sense to establish a Martian base or colony. That presumption may not turn out to be correct.

It could be part of an effort to establish bases or mines on the Martian satellites Phobos and Deimos. If this scenario comes to pass, the most sensible way to reach the Martian surface is to obtain some of the needed resources, such as water and rocket fuel for the return trip, from the Martian satellites.

If most commercial activity in space takes place elsewhere (on the moon or on the near-earth asteroids) or not at all, then the Mars trip will have to be justified on a stand-alone basis, as a primarily symbolic gesture with some scientific by-products. Much of the technology needed for the Mars trip—closed life-support systems, radiation protection, human adaption to extended stays away from gravity—is also applicable to the establishment of colonies in low earth orbit, trips to near-earth asteroids, and to a limited extent to lunar bases. As a result, a trip to Mars could still be coupled to the rest of the space program, even if extension of human activities beyond low earth orbit was rather limited because of the lack of any commercial return.

Thus consideration of the symbolic and technical nature of the trip to Mars is intimately tied up with the question of what we will actually be doing in space in the 21st century. The history of human exploration of the earth suggests that symbolism alone can only sustain a limited exploratory program; while there are always people who are willing to climb every last mountain, finding others who have to stay at home and support the explorers with tax dollars or voluntary contributions is somewhat more difficult once the major milestones have been achieved. Our future activities in space—commercial, military, and scientific—are the focus of the next major section of this book.

* * *

This chapter has brought together a number of diverse threads to address the question: Where will we focus our attention in space in the 21st century? As with many aspects of the space program, our knowledge is much more skimpy than would be desirable. What we do know suggests that the small bodies in the inner solar system—the Martian satellites Phobos and Deimos as well as the near-earth asteroids—are comparatively easy to get to, provide useful sources for materials which can be used in space, and present possible locations for commercial and military use of the space environment.[2]

PART IV

What Will We Be Doing in Space?

Most discussions of the future of the American space program have focused almost exclusively on what destination is most appropriate. The tone was set by the Apollo program, where we seemed most intent on having American astronauts set foot on the moon first and before New Year's Day 1970. The scientific benefits from the lunar expedition were afterthoughts. After all, if astronauts were going to the moon for geopolitical reasons, they might as well do something useful once they got there. This same attitude characterized the space shuttle and space station decisions of the 1970s and 1980s. While both shuttle and station are useful technologies, the decision to go in that particular direction was made because the technology was ready, without seeing what use could be made of it.

The uses of space generally fall into three categories: commercial ("space industrialization"), military, and scientific. Space industrialization is the latter-day equivalent of the spice and gold business of the Golden Age of Exploration of the earth. The communications satellite industry is an excellent—but the only—example of current space industrialization. No other genuine space industries have yet become established as profitable. At the moment, the military's interest in space has generally been confined to the placement of machinery, like Star Wars weaponry, in low earth orbit. The military interest in space may extend beyond low earth orbit in the more distant future, if it should turn out that some particular small, militarily defensible bodies have a particularly high commercial value for space industries.

Space science—the use and exploration of the space environment to extend human knowledge— is one area in which the frontiers are limitless. There is literally no end to the number of projects which can be dreamed up. Space science was begun by a few cosmic ray physicists who could simultaneously do science and help the space program by using cheap Geiger counters to measure the particle environment in near-earth space. It has now blossomed and pervaded many other areas of knowledge: the earth system sciences, astronomy, planetary science, materials science, and the life sciences. Many of the questions space scientists address deal with deep problems: Do planets exist around other stars? How did the universe begin? Is there life elsewhere?

The uses of space, rather than our technical ability to reach particular destinations, should determine where the space program will lead us in the 21st century. We have the ability to go many places: the moon, Mars, the near-earth asteroids, as well as near-earth space. The choice among these destinations should come from answering these questions: Why is there a need for a human presence in some particular part of the solar system, or on some particular solar-system body? Does this location need to be permanently inhabited, occupied by machinery tended by humans, or can remotely operated machines do what needs to be done?

CHAPTER 13

Space Industrialization

Commercial activity in space can cover a wide range, but for the purposes of considering the future of the space program as a whole three reasonably well-defined categories can cover the territory. First, we can put machines—communications satellites and other types of equipment—in space near the earth, without putting human beings there. Second, we can perform experiments and make things by putting both machines and humans in low earth orbit. Third, we can put space industries, people, or both beyond low earth orbit; this form of space industrialization is quite visionary. Since it's impossible to tell whether such industrial activities will be profitable, it's also hard to predict whether people will be needed. Even so, a few sensible comments can be made about which destinations in the solar system seem to make the most sense.

Space industrialization refers to activities where some entity other than NASA or the Department of Defense (DOD) is the ultimate customer for a particular space activity. If space activity is to pay off economically, someone other than a government has to provide a return on the government's investment in space infrastructure. An activity isn't "commercial" simply because the people doing it aren't civil servants on the government payroll. The history of human exploration indicates that some commercial payoff is essential if the exploratory effort is to be sustained; exploration cannot be sustained on symbolism alone. There are a number of space activities in which the payoff is here already, but so far only one small commercial project has required a human presence in space.

While making space profitable might seem most relevant to the largely capitalistic American society, this payoff is as essential for the Soviet Union as it is for the United States. From Sputnik to the present, the civilian side of the Soviet space program continues as a visible way of demonstrating the Soviets' technical equality or superiority in a field which has some military applications. This justification may not remain important to the Soviet bureaucracy for the indefinite future. As a result, the Soviet Union has established the Glavkosmos organization which seeks to obtain a commercial market for its ability to put payloads into space.

Many other countries are showing increasing interest in space industrialization by emphasizing the use of the gravity-free environment of space for experimentation and possible manufacture. These activities were one of seven major themes recently discussed in an ESA planning document.[1] The Japanese module in the projected American space station will also focus largely on microgravity, as did the D-1 ("D" = "Deutschland") spacelab mission which was conducted and paid for by the German government. The Soviet agency Glavkosmos has recently focused its attention on providing opportunities for foreign experimenters to use the microgravity environment on Soviet spacecraft, and has recently signed contracts with France's national space agency CNES and the German company Kayser–Threde GmbH.[2]

A CAUTIOUS INTRODUCTION

In a book of this sort, the main reason for considering potential space activities is to limit as best one can the range of feasible destinations. Since the future scope of space commercialization beyond low earth orbit is uncertain, any discussion of the environment in which that commercialization will take place is equally uncertain. With all this as background, how confident can a reader or an author be in anybody's predictions?

To set the stage, consider what a forecaster 100 years ago would have done in facing a similar dilemma: Where will human beings be living 100 years from now? Someone living in 1890 would not necessarily have been able to forecast the development of the automobile, not to mention the jet plane. Still, it would seem reasonable, living at the time, to predict that the bulk of humanity would continue to live in the temperate zones and tropics, rather than at the poles of the earth. Such a forecast would be right. Forecasting population growth in temperate but underpopulated places such as the American West and the continent of Australia would be both reasonable and, in retrospect, correct. An 1890 futurist would probably not predict our expansion into the suburbs and, not forecasting air conditioning, would probably underestimate the population density in the developed world in hotter climates. Detailed descriptions of future products would no doubt have been way off. It would have been impossible to forecast the development of automobiles, airplanes, and computers.

I'll use the foregoing as a guide, trying to use the general characteristics of various space environments rather than particular products as a guide to reasonable speculation about future space commerce. These general conditions bring us back to the question: Where can space industrialization take place?

AUTOMATED ACTIVITIES IN LOW EARTH ORBIT

The most mature space industries and military activities are those which make use of the ultimate high ground, also described as the orbital vantage point. An orbiting satellite can photograph, receive signals from, or send signals to a large part of the earth's surface. The communications satellite industry is based on this fundamental advantage of a switchboard in space compared to a switchboard on the ground; the range of a transmitter on the highest mountaintop is microscopic when compared to the range of a transmitter on a communications satellite.

The Communications Satellite Industry

The communications satellite industry is a well-established, multi-billion-dollar-per-year business, and it shows every sign of remaining healthy.[3] The basis for this industry is the geosynchronous orbit, a pathway 22,300 miles from the center of the earth, where satellites orbit the earth every 24 hours, following the earth's rotation precisely. Thus the satellite keeps pace with the earth's rotation, and remains fixed in the sky from the viewpoint of earthbound observers who wish to receive signals from it. While the limited number of "parking spaces" in geosynchronous orbit may limit some possibilities for expansion of this industry, there are ways around this apparent limitation on communications channels and there are a number of other markets, particularly outside the developed countries, which are opening up. This industry has changed our lives in many small but significant ways. Global networks of cash machines make travel easier. Intercontinental telephone calls are commonplace. Electronic mail networks make it easy for me to talk to co-workers in Toulouse, France; Kiel, West Germany; and one who normally is in

Seattle, Washington, but on his one-year stint in Australia can check his Seattle electronic mailbox remotely, even though the "mailbox" is halfway around the world! This global interconnectivity has its disadvantages, as I found recently; some of my collaborators really expected me to check my electronic mailbox every day that I was on vacation.

Television network news departments, in their late-night shows, have used the satellite capabilities to initiate dialogues between various countries. Ted Koppel's "Nightline" program has been able to engineer several confrontations or dialogues between opposing groups like South African supporters of *apartheid* and American opponents of the policy, and the Nicaraguan government and American *contra* supporters (and, as I recall, *contra* officials). Even more enlightening were the "Capitol to Capitol" dialogues between American congressmen and Soviet mid-level bureaucrats. These live programs seemed to promote a climate of understanding and even trust after the interchange.

Variations on a Theme

The orbital vantage point which the communications satellites use has produced a number of commercial or quasi-commercial activities which are closely related to it in that they make use of a satellite's ability to receive easily and quickly light or signals from all over the globe and do something with them. Weather satellites photograph cloud patterns, receiving light signals from the earth, and retransmit them as radio signals to a host of private and public receivers. Satellite-based rescue systems receive emergency radio signals and retransmit them to appropriate rescue stations. And the beat goes on.

Most familiar are the global photographs of weather patterns, a familiar feature of evening television weather broadcasts. The provision of these pictures is "commercial" in a rather vague way since Americans are used to getting free weather forecasts from the National Weather Service; the idea of paying for accurate weather information, no matter how useful it might be, seems rather alien. But prevalence of private, profit-making companies in the weather forecasting business ("Accu-Weather," based in State College, Pennsylvania, provides weather information to many radio stations in the Northeast) shows that there's a definite market for these services. Thus the weather forecasting business could possibly be considered "space industrialization" in some sense, even though the space side of weather forecasting does not involve the sale of space services to private customers.

Another potentially profitable use of space services, closely related to weather forecasting, is the use of pictures of the earth's surface for mapping, urban planning, news reporting, and other commercial activities. The commercial potential of this activity was considerably oversold in the 1970s, fueling some expectations for a nonexistent business. It didn't help matters that the American satellite which was the candidate for commercialization, Landsat, produced images which were considerably lower in quality than the technology permitted. Military interests prevailed, and Landsat's capabilities were intentionally degraded to disguise the greater capabilities of our reconaissance satellites. This passion for secrecy was misguided; as journalist and space analyst William Burrows has shown, it's quite simple to deduce the rough capabilities of our spy satellites from general principles.[4]

True commercialization of the space imaging business may come through the French corporation SPOT Images. SPOT (Systeme Probatoire pour l'Observation de la Terre, or, in English, Earth Observation System) is a French satellite which can rapidly return images of areas of the earth with a resolution of 30 feet, three times better than Landsat. A few days after the Chernobyl nuclear plant burned up, pictures of the disaster, taken by SPOT, appeared on TV network news.

A still more recent, and rather ingenious, use of the orbital vantage point is being run by Geostar, a commercial satellite which sells the service of locating a client's properties on the surface of the earth. Geostar has 42 clients; 35 of them are trucking companies that want to keep track of their trucks' location, cargo, and routes. The Coast Guard is also a trial customer, and Geostar hopes to sell its services to the aviation industry too.

Currently, Geostar uses the ground-based Loran system to locate their customers' trucks. Radio transmitters in the trucks signal their locations to Geostar transponders in orbit, and the locations are retransmitted to company headquarters. By the early 1990s Geostar expects to use satellite navigation techniques to locate its customers' vehicles, allowing the company to extend its geographical coverage beyond the area covered by the Loran system. While Loran provides good coverage of the continental United States, it only provides positions within half a mile (1 km) and satellite navigation can be considerably more accurate.[5]

Last let me mention the "Dick Tracy radio"—a prediction in my previous book which has partly panned out in the short span of two years.[6] An idea which has been around the space business for a while comes from Chester Gould's Dick Tracy comic strip. Detective Dick Tracy, who first appeared in comics many decades ago, had a wrist radio which he could use to communicate with headquarters. In the 1960s technology caught up with the comics, and the wrist radio was upgraded to a TV. While the Dick Tracy radio in the comic strips had a limited range, the space-age descendant of this idea is a portable radio set which could tie in to the worldwide phone system by directly communicating with a communications satellite.

We don't yet have Dick Tracy radios, but we have something that's one step away from them—the satellite-triggered, nationwide beeper. For a price, customers can buy a beeper which is triggered not by some local, ground-based radio transmitter, but by a signal from an orbiting communications satellite. A jet-setting company president can be summoned to the nearest phone from wherever he or she may be. With cellular phones making it possible for people in cars, on trains, and on airplanes to tap into the national phone network, the communications potential of the Dick Tracy radio is practically here.

It seems reasonably clear that in the very early part of the 21st century, the amount of commercial machinery in low earth or geosynchronous orbit will expand considerably. However, putting machines in orbits near the earth offers little incentive for humans to expand our sphere of influence beyond low earth orbit, or even to establish a significant presence there. All existing and most forseeable activity which uses the orbital vantage point in this way only requires robot eyes and ears in space rather than people. Arthur Clarke's initial vision of communications satellites and some science fiction novels of the 1950s had people in geosynchronous orbit—telephone operators or TV news anchors sitting in the global "catbird seat."[7] Computer technology caught up with this idea, and people are no longer needed in orbit.

Promoters of space settlement have responded that a manned repair facility or two will be required if there is much commercial activity in low earth orbit. However, it remains to be demonstrated that there is a substantial market for cost-effective in-orbit repair of satellites. In November 1984, NASA proudly announced that it had rescued two communications satellites which had failed to reach geosynchronous orbit and brought them back to earth. While this was a valuable demonstration of a neat new technology, in this instance I doubt that the rescue would have been cost-effective if NASA had charged the communications satellite companies for the full $500 million cost of a space shuttle flight. The $500 million might well cover the cost of launching two brand new satellites. While space satellites may become sufficiently expensive that repairing them is cheaper than replacing them, the consumer electronics market is heading in the other direction. In space, the Soviet Union's approach has been to build lots of cheap satellites and simply replace them when they fail. To justify a more expansionist view, we must turn to other commercial activities.

USING ASTRONAUTS IN LOW EARTH ORBIT

Still at the exploratory stage are some ideas in which two aspects of the space environment may provide some commercial opportunities. The environment inside an orbiting space shuttle or space station is a weightless environment, more accurately described as a microgravity environment. A decade or so ago, some forecasters stimulated the imaginations of scientists and managers alike by predicting that many new valuable materials could be developed in microgravity. These products are further from reality now than they were a decade ago, primarily because the promise of low launching costs from the space shuttle has not materialized. At this time, orbiting spacecraft are generally seen as laboratories in which an experimenter can turn gravity off and see what happens to a particular process like crystal growth or the solidification of a cast-iron engine block.

Space Manufacturing

In the 1970s there was considerable discussion of space manufacturing. The name of NASA's program in this area, Materials Processing in Space, provided clear indications that what was visualized was vast manufacturing facilities in orbit which would produce valuable materials. While there has been one successful venture along this line, progress has been distinctly limited.

Dr. John Vanderhoff, a scientist at Lehigh University, worked with NASA's Marshall Space Flight Center to develop an apparatus called the Monodisperse Latex Reactor (MLR), the purpose of which was to produce millions of tiny plastic balls, all perfectly spherical and all of the same diameter. ("Monodisperse" means "identical size.") These spheres are shown here in the photograph along with a collection of spheres made in the same apparatus when it was on the ground. While the lighting conditions are slightly different, it is evident that the spheres made on the ground are not all round, nor are they of uniform size. The National

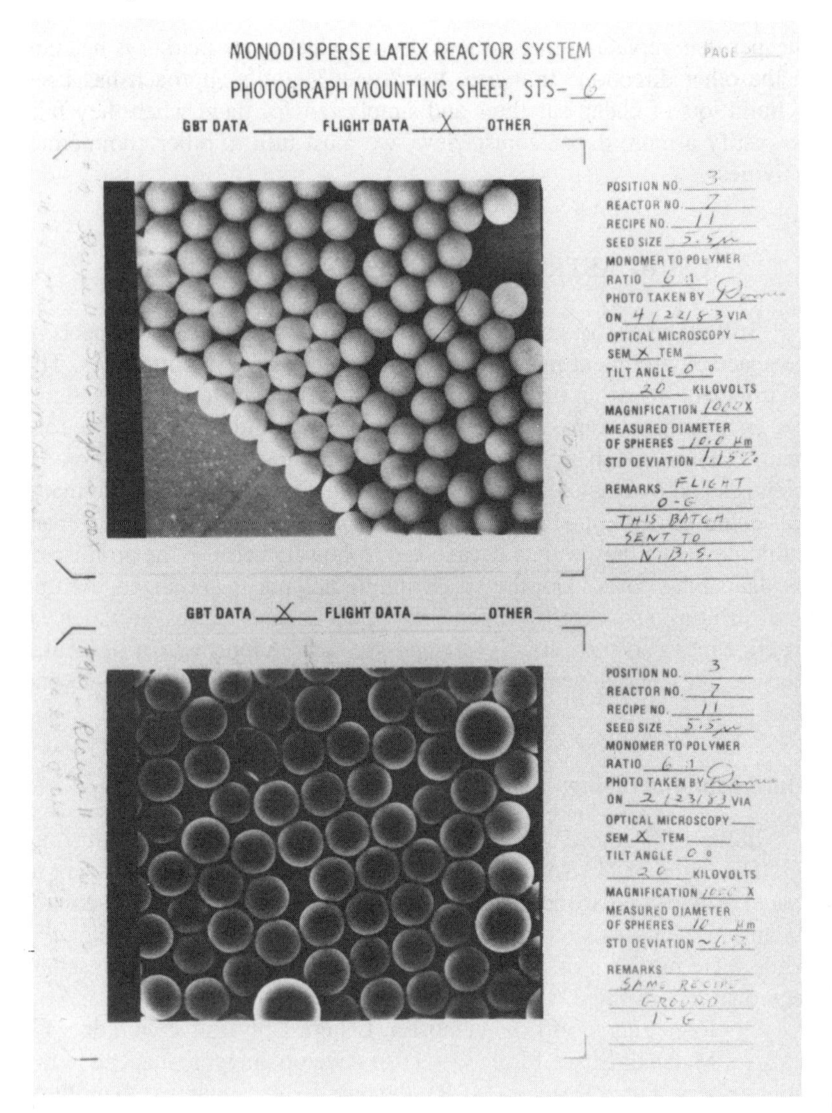
MONODISPERSE LATEX REACTOR SYSTEM

PHOTOGRAPH MOUNTING SHEET, STS-6

GBT DATA ______ FLIGHT DATA X OTHER ______

POSITION NO. 3
REACTOR NO. 7
RECIPE NO. 11
SEED SIZE 5.5μ
MONOMER TO POLYMER RATIO 6:1
PHOTO TAKEN BY
ON 4/22/83 VIA
OPTICAL MICROSCOPY ______
SEM X TEM ______
TILT ANGLE 0°
20 KILOVOLTS
MAGNIFICATION 1000X
MEASURED DIAMETER OF SPHERES 10.0 μm
STD DEVIATION 1.15%
REMARKS FLIGHT 0-6 THIS BATCH SENT TO N.B.S.

GBT DATA X FLIGHT DATA ______ OTHER ______

POSITION NO. 3
REACTOR NO. 7
RECIPE NO. 11
SEED SIZE 5.5μ
MONOMER TO POLYMER RATIO 6:1
PHOTO TAKEN BY
ON 2/23/83 VIA
OPTICAL MICROSCOPY ______
SEM X TEM ______
TILT ANGLE 0°
20 KILOVOLTS
MAGNIFICATION 1000 X
MEASURED DIAMETER OF SPHERES 10 μm
STD DEVIATION ~6%
REMARKS SAME RECIPE GROUND 1-6

The top photo shows plastic spheres made in space. All of them are round, and they are almost the same size; measurements show that two thirds of these spheres are within 1.15% of the diameter they are supposed to be (10 micrometers). The bottom photo shows similar spheres made on earth. Some of the spheres are not round, and some are considerably larger than the others.

Bureau of Standards has certified that the diameters of the spheres made in space are quite uniform. They are currently being sold at $384 for a 5-milliliter vial containing 15 million spheres.[8]

What in the world would anyone want to do with millions of identical tiny plastic balls? A number of potential customers exist. People who manufacture toner for office copiers want to make a fine powder with particles of uniform, predictable size; an easy way to measure the particle size precisely is to calibrate a microscope using one of these tiny balls. Sales of the balls made on the single space shuttle flight in 1983 have not been brisk, and even if the whole lot is sold, NASA will only realize $250,000. As a start for space industries, this is rather modest. Apparently, most of the market for plastic microspheres is for bigger balls (the ones on sale now are 10 micrometers in diameter). NASA mentions a potential market of $1 billion per year for this industry.[9] The MLR is, so far, the only example of a salable product made in space.

Considerable publicity and considerable investment of time and money on the part of McDonnell Douglas in another venture has not led to any tangible results. McDonnell's idea was that separation of medicines could be done considerably better in orbit because a number of effects which are caused by gravity don't exist in microgravity. The shortage of flight opportunities caused McDonnell's partner, Ortho Pharmaceuticals, to drop out of the venture, and McDonnell Douglas has now joined the growing queue of people with ideas and equipment who await launch opportunities.

Scientists and engineers face several basic difficulties in asking the question, "What can we make in space that can be sold at a profit?" The first problem is that the tangible and intangible cost of doing business in space is considerably higher than the cost of similar technologically innovative activities on the ground. Ray A. Williamson of Congress's Office of Technology Assessment has identified a number of unique problems which inhibit commercial innovation in space.[10] Because various governments control the means of access to space, any investor is held hostage to space shuttle launch schedules and variable government budgets. Spacecraft casualties are not just American; Russian and Chinese rockets blow up too, and investors whose payloads are on rockets that explode either lose their investment entirely or find that the cost of insuring a particular rocket payload is exorbitantly high. While no innovative industry is risk-free, the lack of control of access to one's laboratory or factory is something which is unique to space industries.

The tangible costs of getting into space are also very high when compared with other industries. It costs several thousand dollars per pound to put a piece of equipment in orbit. If establishing corporate research labs cost that much, there would be very few—if any—corporate research labs. Consider a particularly simple example, in the context of space manufacturing. If it costs $1 million to launch a furnace that weighs 100 pounds or so, then even if that furnace generates 100 pounds of product you have to sell the product for $10,000 per pound in order to make any money. Furthermore, that simple calculation doesn't allow for any mistakes or for any exploratory research, and also makes the unlikely presumption that the product weighs about the same as the furnace that created it. Space manufacturing is clearly limited to very high-value products.

But high-value products exist, and risks and uncertainties are inherent in any technologically innovative industry. It would be premature to conclude, on the basis of the cautionary remarks above and on the basis of the limited success, so far, of the space manufacturing efforts, that there is no commercial future to microgravity.

There is yet an additional barrier to space manufacturing: relative ignorance of the actual opportunities presented by the microgravity environment. In the past, new materials have been developed only after scientists have developed a basic understanding of the physical and chemical processes which take place when those materials are formed. In order to make plastics, for example, an engineer must understand how small chemical units stick together to form the huge molecules which are characteristic of plastics. Scientists can guess what materials will do in a microgravity environment, and much of what was written in the 1970s was based on just such speculation. But only actual experimentation can show scientists just what does happen and how different types of materials can be synthesized in space.

The Space Environment as a Materials Research Laboratory

One way to find out what happens in space is to send some chemists and physicists up into space and simply let them fiddle around. A better way, which has been exploited with increasing frequency in recent years, is to recognize that the space environment can be used as a special type of research laboratory, one in which gravity can be turned off. In this sense, a space laboratory is quite analogous to, for example, the National Magnet Laboratory, where scientists can bring their samples and place them in very strong magnetic fields. An example will give some of the flavor of what one can do in a space laboratory.

One of the most unlikely participants in any space laboratory experiment is Deere and Company, the prime manufacturer of farm tractors in the United States. Farm tractors in space? The idea sounds ludicrous, at first. Farm tractors are big, heavy behemoths, which grunt their way loudly through the fields. Their wheels stand taller than most humans. Deere makes a lot of them and paints them all bright green. I find it hard to imagine anything quite so far from the space program. Yet Deere has signed a Technical Exchange Agreement with NASA.

One quarter of the weight of a typical farm tractor is the cast iron in the engine block. Deere engineers realized, after some discussion, that the development of graphite flakes in the cast iron has a great effect on its properties. How do these graphite flakes form? Gravity might have an effect on the formation of these flakes, because any temperature differences which exist in the cooling, solidifying cast iron will cause hot liquid to rise and cool liquid to sink, mixing the stuff up. The only way to stop these rising and sinking motions is to turn gravity off, and so Deere turned to NASA. There was never any intent of casting engine blocks in space; the motivation was developing better understanding which could then lead to a better product and more profits.

Space flights are expensive, but NASA maintains a number of other facilities in which a microgravity environment can be created for a short time. Deere used KC-135 and F-104 aircraft which, when flying on a special path, can produce a low-gravity environment in their interiors. Deere built a furnace and the automated instrumentation which could follow the cooling process, and flew this equipment on NASA's airplanes. James Graham of Deere, writing in 1986, reported that some differences were indeed found when cast iron solidified in microgravity, and the company does plan future research with NASA.[11]

Several other companies have participated in similar ventures in which space is used as an applied science laboratory. The 3M company has studied crystal growth in orbit. GTE has studied how arcs in arc lights (the kind that are used to illuminate ballparks at night) work in the absence of gravity. Other studies in microgravity are more fundamental and, apparently, more distant from direct profitability. A recent blue-ribbon panel of the National Academy of Sciences concluded that these fundamental studies are a necessary foundation which should precede any realistic assessment of the potential for space manufacturing.[12]

Ultra-High-Vacuum Research

Another aspect of the space environment, even less well exploited, is the ready availability of a very good vacuum. Scientists and engineers can create reasonably good vacuums on earth by pumping the air out of a cavity and chasing down the leaks which let the air seep back in, but there are practical limits to the quality of the vacuum which can be obtained on the ground. There is no air in space—it is a good vacuum indeed. A number of industrial processes work only in a good vacuum. Some electronic devices, for example, are built up by the precise deposition of individual atoms on the surface of crystalline silicon.

At the altitudes where the space shuttle flies, there's still a bit of terrestrial atmosphere left, but there is an ingenious way to get rid of it. Alex Ignatiev of the University of Houston has explored ways of using the Space Ultravacuum Facility (SURF), scheduled for a shuttle flight in 1991. SURF is a 10-foot-diameter shield which will be deployed into space on the shuttle's remote manipulator arm. SURF will be oriented perpendicular to the direction that the shuttle is flying in, and in its wake the pressure will be 10^{-14} Torr, 1000–10,000 times better than the best terrestrial vacuum. Ignatiev's particular interest is in using a technique called molecular beam epitaxy (MBE) which can build up layers of atomic material which are much more defect-free than those synthesized in the best terrestrial vacuums. If this scheme works, it may be possible to fabricate computer chips from materials which can't be used now because it is impossible to grow sufficiently good microscopic structures.[13]

For those who seek to justify the human presence in space, these uses of the space environment offer some hope which is lacking in the communications satellite industry. These newer commercial activities require human experimenters, unlike space communications which only require machines in orbit. Unfortunately, it's also true that the work which has been done so far has not yet generated commercial activity and return on investment of a scale similar to communications. A space materials industry does not yet exist, even as a service organization which can provide a lab in which gravity is turned off. Because of the long lead times and substantial uncertainties associated with space activities, it would be premature to conclude that the space materials industry will never exist. Indeed, the sort of applied studies conducted by companies like Deere can continue indefinitely, though whether these companies would be willing to pay the full cost of access to space in connection with such studies remains to be seen.

Space enthusiasts have seized on the potential for microgravity work, as they seized on the solar power satellites of a decade ago, as one way of justifying the human presence in space. Interest in solar power satellites waned when Arab oil became cheap again, and solar power satellites probably wouldn't make sense even in a world where solar power was our main source of electricity.[14] The microgravity sciences are broader in scope, and it's not likely that the collapse of one particular market will result in a collapse of interest in microgravity. However, it's premature to say that microgravity science will make space settlement pay for itself.

Yet the possibility that microgravity (or something else) may pay off in the distant future has caused all of the world's major economic powers to sustain an interest in a space program, and the interest in microgravity appears to be increasing as time goes on. It would be as wrong to discount the potential of microgravity prematurely as it would be to sell microgravity as a sure bet for space commercialization. Financial writer Gail Bronson, writing in *FORBES* magazine, concluded that since NASA had spent five years trying to drum up industrial interest in space shuttle opportunities and had had little success, space commercialization was, in the words used in the title of her article, "mission irrelevant."[15] Such a focus on the short term is regrettable; we should not write off microgravity as a possible major industry in the 21st century.

BEYOND LOW ORBIT

If the prospects for microgravity are uncertain, the prospects for a significant amount of self-supporting human activity outside of low earth orbit are more uncertain still. One possible justification for human habitation outside low earth orbit is the possible extraction of high-value materials from certain types of asteroids. A scenario which is still speculative but a little closer to hand, both economically and technologically, postulates extensive human settlement of low earth orbit and then justifies bases beyond low earth orbit as ways of supplying necessities of life to the low orbit settlements. Neither of these scenarios is particularly certain at the present time, but both suggest—as forewarned in Chapter 12—that small bodies like near-earth asteorids, Martian moonlets, or both are the most likely sites for settlement of the inner solar system in the 21st century.

Mining

Large-scale, sometimes temporary migrations of human beings to improbable, inhospitable places can be justified if there's an easily extractable, highly valuable commodity there. Gold was the magnet which drew Spanish explorers to the Americas in the 1500s, Americans to California in 1849, and Americans and Canadians to the Klondike in the 1890s. One of the most exciting proposals for space industrialization is based on a similar substance, a residue of heavy elements which can in principle be extracted from carbonaceous types of asteroids.

This proposal of asteroid mining is quite exciting, whether or not it ultimately bears fruit. At last someone has conceived of something which can be made or extracted in space, can be exported to earth for a price which is higher than the cost of carting materials to orbit, and is sufficiently rare on earth that the likelihood of terrestrial competition catching up with a potential space industry is quite low. It's true that we know so little about asteroid mining that no reasonable person could come up with a prospectus which would set out the economics in enough detail to satisfy even the most visionary of bankers. However, there are a number of aspects of this idea which strike me quite favorably:

(1) We know that there is a sustained market for gold, platinum, and other heavy metals. Other space commercialization efforts have produced products (such as the tiny plastic spheres) for which the market is not established and may be saturated by one or a few space missions.

(2) The raw materials are obtained in space and need not be carried up from the ground. Space industries (such as crystal growing and drug separation) in which stuff is carted up from the ground, processed in the microgravity environment, and taken back down again face the freight bill of several thousand dollars per pound for whatever is produced. Indeed, it's conceivable that over the long run the mining equipment and the rocketry needed to reach this asteroid and bring metals back from it could weigh less—or not much more—than the metals which are produced, especially if solar sails are used.

(3) In this area of space commercialization, it is very unlikely that human ingenuity can find ways to mimic the advantages of the space environment as has happened with other fledgling space industries. The communications satellite industry had to face competition from fiber optics cables. McDonnell Douglas's drug separation apparatus in space may have only stimulated further ground-based research in separation techniques and not made money for McDonnell Douglas.[16] There's plenty of incentive to find gold mines on earth without any stimulus from space industries; competition for asteroid mining already exists.

If precious-metal extraction is to become a significant space industry in the 21st century, the asteroids, Phobos, and Deimos grow in importance relative to other possible destinations. It's true that the economics of asteroid mining are very uncertain. We do, however, know that the platinum-group metals are present in carbonaceous chondrite meteorites, which are presumably similar to an important subclass of asteroids. In addition, the iron and nickel asteroids could also play a role as sources of raw materials for future spacecraft.

Supporting Infrastructure for Low Orbit Settlements

I turn now to a possible space industry which is a little different from those discussed so far. Since my purpose is to seek human activities which can make space settlement as a whole self-supporting, I have so far discussed "space industrialization" activities which have sold products or services directly to customers outside the aerospace industry or the Federal government. It's possible that space settlements in low earth orbit will become self-supporting and will require supply from other, secondary space industries which will justify an extension of human presence, beyond low orbit, in the form of either remotely operated robot mines or settlements.

Space settlements in low earth orbit will need water and other life necessities in order to exist. Where could this water be obtained from? The surface of the earth contains plenty of water, but considering the rocket fuel needed to lift this stuff off the earth's surface, it's very far away from low orbit. In principle, virtually any other destination, be it the lunar surface, Martian surface, Phobos, Deimos, or a near-earth asteroid, is closer. We know there's water on Mars (but is there enough of it?). Phobos, Deimos, or the near-earth asteroids may also contain water, and they are easier to reach and return from.

A current example of such a secondary space industry which may become self-supporting is the private launch vehicle industry. With space on the space shuttle suddenly very scarce in the late 1980s and early 1990s, a number of companies have sprung up which launch communications satellites for paying customers, for a price which does not include a subsidy. These companies compete with each other and with foreign-government-sponsored space organizations. Partly because the foreign competitors may be subsidized, American companies in this industry may have a tough time of it. Nevertheless, it seems clear that there is ample opportunity for at least one or two companies to keep going for the forseeable future.

Space Industries on Mars?

While it's possible to speculate endlessly about other commercial activites in space, the above discussion suggests a focus which can limit the speculation: What about commercial activities on Mars? All of the discussion so far suggests that economically self-supporting activites beyond low earth orbit are most likely to occur on Phobos, Deimos, or an asteroid—not on the surface of Mars. But Mars has been widely mentioned in various reports as the next site for a human outpost. Is the proposed outpost on Mars merely a symbol of human achievement, or is it intended to become the nucleus for a self-supporting colony? To put it very crassly, is there any way to make money on the surface of Mars?

Mars has been little discussed as a source of nonterrestrial materials. Mining Martian materials, first mentioned in a 1981 Case for Mars conference, has been conspicuous by its absence from authoritative discussions of extraterrestrial resources.[17] Most writers regard the moon as easier to get to in terms of travel time and launch window frequency. While there is water on the surface of Mars, and while a Martian colony would probably use it, it's not likely that it would make much sense to export water to low earth orbit. While I could certainly be wrong, on general grounds the abundance of sand and the weakness of plate tectonic processes does not encourage the idea that concentrations of rare materials will exist in special locations on the Martian surface.

However, even in the market of the late 20th century, selling services is as important as selling goods. Let me mention some rather speculative—I know some readers would say ridiculous—possibilities in an effort to spur further thinking. What is there about Mars that's both different from the earth and different from a space station in low orbit? Mars has gravity, but the gravity is weak. Perhaps people with some diseases like arthritis could avoid considerable pain and live longer, more productive lives if they were to live on the surface of Mars, weighing about 40 % as much as they weighed on the earth. Perhaps sex in such an environment would be astonishingly satisfying, making Mars a popular resort for the very rich. Perhaps athletic or dancing events in this low gravity would be so spectacular that it would be worth sending the performers to the red planet to make videos. Perhaps the Martian landscape, with its spectacular volcanic mountains up to 25 miles high would be a major tourist destination.

I really doubt that 21st century Mars will become an elite resort like Monaco, but I'm trying to think of something for Martian colonists to do which makes commercial sense and is not totally ridiculous. There is an acute shortage of ideas for commercializing Mars. This shortage may simply reflect a shortage of imagination; if not, the absence of any conceivable commercial return may become a serious obstacle to colonizing Mars.

* * *

The commercial side of space definitely exists, in the form of the communications satellite industry and other variations on this theme which take advantage of the orbital vantage point. Expansion of these types of activities will undoubtedly occur in the 21st century. However, machines rather than people can occupy the orbital vantage point and run communications networks very successfully. This type of space industry is not likely to justify space settlements in low earth orbit. The unique properties of the space environment—most particularly microgravity—have suggested to many that humans in low earth orbit could justify their existence by conducting applied research or, possibly, by manufacturing unique materials.

While prospects for a genuine increase in new types of space commercialization are still quite far off, every one of the major economic powers on earth has decided to venture into space. Apparently, if a nation wants to be considered a major player in the late 20th-century world, some attention to space, and in particular to space industrialization, is called for. Space is clearly not just a soldiers' and scientists' playground.

Note added in proof. In late 1988, McDonnell Douglas abandoned the medicine separating venture mentioned on pages 253 and 258.

CHAPTER 14

The Military Uses of Outer Space

Occasionally, nations have sought to establish human settlements in various locations because of their potential military value. One of the best examples of such a settlement is the British garrison at the Rock of Gibraltar, and the western entrance to the Mediterranean Sea, south of Spain. When it was established, the British had no designs on taking over the Iberian peninsula; rather, this maritime power sought to establish rights of access to the Mediterranean Sea by establishing a fort in a strategic location. Another example of such bases is the American base at Diego Garcia, a remote islet in the southern part of the Indian Ocean. Diego Garcia is not seen as a site for American commerce or mining; rather, it is a convenient place for refueling and communications.

With the prospects for commercial activity in space somewhat uncertain, a natural question to ask is whether there are any other activities in space which could, by themselves, justify the extension of the human presence beyond the surface of our planet. Especially in the "Star Wars" era, military activities come readily to mind, although they make many space enthusiasts and space scientists rather uncomfortable. Two types of military activities in space are conceivable: those which are focused on other military activities taking place on the surface of the earth, and those which accompany space settlements.

Present and conceivable future military activities in space range from the established and accepted reconnaissance satellites through futuristic concepts like former President Reagan's "Star Wars" dream and even beyond. These projects focus on the earth, do not place hardware elsewhere in the solar system, and only require the presence of people in space to the extent that repair and refurbishment of some of the more sophisticated pieces of machinery might make economic sense. There is no conceivable need to establish military bases on the moon, and in fact the 1967 Outer Space Treaty explicitly forbids military bases or territorial claims on the moon or on any other celestial body. Thus military needs cannot be used to justify some kind of space settlement scenario.

If settlements are established at various locations within the solar system, territorial or military considerations may not remain forever confined to the surface of the earth in spite of the 1967 Outer Space Treaty. The extreme vulnerability of most space settlements will make the military developments beyond the earth, if any, rather unlike those we are familiar with on the earth's surface. There are a few locations, most notably Phobos and Deimos, which are conceivably defensible in a military sense and thus could be claimed as the territory of a particular nation.

PASSIVE MILITARY SATELLITES IN LOW ORBIT

While some bemoan the "militarization of space" which seems to have taken place in the 1980s, the military presence in space is as old as the space age itself. In the Soviet Union, there is no distinction between military and civilian space programs. While such a distinction supposedly exists in the United States, Wernher von Braun's group working under the auspices of the United States Army launched the first American satellite in January 1958. In the 1970s, some crucial aspects of the space shuttle design were determined by military needs.

Both the United States and the Soviet Union have devoted virtually all of their military machinery to passive activities like reconnaissance and communications. Political scientist Gerald Steinberg has drawn a distinction between such passive systems and active space weaponry which includes antisatellite weapons systems, missile defense, and the like.[1] The primary function of these passive systems is to support military operations, rather broadly defined, on the ground. As with civilian communications satellites, the primary advantage of being in space is the orbital vantage point which allows a particular satellite to be in contact with, or see, a much larger part of the earth's surface than is possible from airplanes.

Reconnaissance satellites have played a unique role in the global geopolitical situation. We probably would not have any arms control agreements at all if these satellites did not exist, because without them verification of compliance by the other side would be impossible. These are an excellent example of a nonthreatening, passive weapons system; it is likely that the competition between the superpowers would be much more intense if these satellites did not exist.

In his excellent overview of the history and capabilities of space satellite systems, political scientist and journalist William Burrows has painted a fair and accurate picture of what the cold war was like in the 1950s before the advent of spy satellites.[2] From the American perspective, the military threat of the Soviet Union was clouded in secrecy, hidden in a closed society which declared vast regions of Siberian tundra off limits to foreigners. Three technological surprises—the Soviet atomic bomb, their hydrogen bomb, and the launch of Sputnik—combined to create an image of a mysterious, powerful enemy whose capabilities could only be imagined. American military officers produced enormously inflated estimates of Soviet military capabilities.

Sometimes the Soviet Union came up with clever ways to make its own people and the Americans think it was more capable than it really was. Colonel Charles E. Taylor of the United States Air Force, stationed in Moscow in 1955, watched the annual Soviet air parade. Thinking that the entire Soviet air force was passing over the airport for review, he was stunned to see wave after wave of Bison intercontinental bombers passing overhead. Taylor counted the number of planes he saw, and American analysts used his counts to predict that hundreds of Bison bombers would be available by the end of the decade. It turned out that each wave of Bison bombers consisted of the same planes, which had simply made a wide circle and flown over the airport again. Only 18 Bison bombers were airworthy that day,[3] a great contrast with the American estimate of nearly 100. Such uncertainties seem inconceivable in an era where each side knows exactly how many bombers, missiles, and warheads the other side has.

The capabilities of reconnaissance satellites are indeed astounding. One type of satellite takes pictures covering large geographical areas in order to pinpoint those locations which would benefit from more careful watching. Another type of satellite takes the close-up shots in which significant details can be distinguished. While the sharpness, or resolution, of the spy satellite photographs is supposedly a closely guarded secret, a number of sources in the open literature have quoted estimates of 2–6 inches as the size of the smallest feature that one of these satellites can photograph.[4] Pictures of the mirrors used instead of lenses in these satellites have been published,[5] and estimates based on the general size of these things confirm the 2-to-6-inch figure.[6]

If we can distinguish features which are 2–6 inches across, photo interpreters can precisely identify militarily important hardware like rockets, artillery, aircraft, and even nuclear weapons components. Open congressional testimony indicates that photos with a resolution of only 3 feet are sufficient to identify an aircraft type.[7] Thus, in a space satellite era, we would no longer have the spectacle of American analysts interpreting a fleet of 18 bombers as being hundreds strong because a trick was played at a Soviet air show.

In the future, military satellites in low earth orbit will become increasingly sophisticated and capable of determining what the other side is doing. The Soviet Union has developed one type of spy satellite, launched in low orbit, and they can launch it very frequently. Their strong point is that they can respond quickly to the loss of one of their satellites, or to a need for a quick reconnaissance mission. Burrows cites an incident where the Soviet Cosmos 1221 satellite was launched right after an American military exercise in the Egyptian desert began. Cosmos 1221's orbit carried it directly over Egypt at the announced beginning of the exercise.

The American approach to spy satellites is somewhat different, appropriate to a technologically more sophisticated country with launch opportunities which were limited even before *Challenger*. American spy satellites are very fancy gadgets which can do almost anything. Once in orbit, their orbits are inflexible, so they can't respond to strategic situations, but they stay up for a long time. An indication of their sophistication is their cost, which Burrows estimates at $100–300 million.[8]

Looking ahead, it is likely that there will be more and more passive systems like spy satellites in low earth orbit, and the satellites will become increasingly sophisticated. In the near term, the most serious gap between what the military has and what it wants is timing. When something like the Chernobyl nuclear disaster occurs, it would be desirable to obtain an image of that area quickly. With several spy satellites in orbit, one could be redirected in order to obtain such a picture on very short notice. Since we had only one, our first pictures of Chernobyl came from the French SPOT satellite. Other capabilities of passive satellites include locating ships and airplanes with high precision, intercepting the other side's internal communications, monitoring the globe for evidence of clandestine nuclear tests, and performing all of the technological tricks which are part of the military trade of peacekeeping in modern times.

Peacekeeping? Yes, I did mean peacekeeping. For the most part, these passive military activities have been accepted by both sides, and by various political factions on both sides, as being legitimate ways of managing superpower conflicts. Information from spy satellites plays an important role in reducing nuclear proliferation, verifying compliance with the few arms control treaties that do exist, and in general in moderating the tone of international disputes. There are, however, military activities in low orbit which are considerably more controversial.

ACTIVE WEAPONS SYSTEMS IN LOW ORBIT

One of the most significant changes in the American approach to space during the Reagan Administration concerned "active" weapons systems, facilities in low orbit which don't simply enhance ground-based military capabilities but rather go beyond that and threaten to destroy targets in orbit or missiles en route to their destinations. The most widely publicized of these active weapons systems is the proposed—or more accurately dreamed of—missile defense system which has recently become far more visible as a result of certain actions of the Reagan Administration. The changing American attitude about antisatellite weapons systems may be as significant a Reagan legacy as the Strategic Defense Initiative (SDI) program.

Star Wars

Few space proposals since Kennedy's lunar landing speech have stirred passions as much as the Reagan Administration's Strategic Defense Initiative (SDI). Much of the debate has been rather fruitless because there are some widespread misperceptions as to what SDI is and what SDI is not. Let me deal with some of the misconceptions first.

SDI is not a magic antidote to mutual assured destruction. Americans since the time of Harry Truman in the late 1940s have been distinctly uncomfortable because both nuclear superpowers have the capability to destroy the other. What has happened during the 45 years of general global peace since 1945 is that both superpowers have been sufficiently awed by the other's nuclear arsenal that regional conflicts like those in the Middle East, Vietnam, Afghanistan, and Angola have not escalated into Armageddon. While this state of affairs was recognized with the name of mutual assured destruction and blessed with the singularly appropriate acronym "MAD," no one likes it.

However, escaping from MAD is not as simple as one would like. The simple fact is that the detonation of a few dozen nuclear bombs over randomly selected targets in the United States can bring our society to its knees.[9] The Soviet Union has roughly 10,000 nuclear warheads mounted on a variety of delivery vehicles. With 10,000 Soviet warheads available, if only 1 % of these warheads were to get through our defensive scheme, we've had it. No conceivable defensive scheme, including SDI, can be 99 % effective. Star Wars, visualized as a shield, won't work.

Note added in proof. The Bush Administration has accepted this fact.

The idea of missile defense is not new. Missile defense schemes have been discussed since the 1950s. The futuristic laser technologies which were at one time the centerpiece of the SDI program have been under study at least since the early 1970s. The principal result of President Reagan's "Star Wars" speech, in 1983, was to make these technologies highly visible, considerably better funded, and to raise the national consciousness about the remote possibility of an impenetrable nuclear shield which could, in the president's words at the time, make ballistic missiles impotent and obsolete.

SDI: Five Years Later

The result of five years of debate on SDI has been considerable confusion regarding what SDI is. SDI proponents cannot understand that anyone would be opposed to the idea of an impenetrable shield, and accuse SDI opponents of actually liking mutual assured destruction. However, the idea of an impenetrable shield is a long way off, and any realistic debate about SDI has to be about the kind of system we can implement or research we can accomplish in the relatively near future. SDI opponents have tended to criticize the whole SDI program instead of limiting their attacks to its more extreme aspects. Politicians take stands in favor of or against "Star Wars" without specifying what is meant by "Star Wars." Does "Star Wars" mean research, deployment of a relatively primitive system in the 1990s, or eventual deployment of a very sophisticated and expensive system at some future date?

The result of five years of research on SDI has been an increased focus on traditional technologies at the expense of some of the more futuristic ones, and a realization that many technological breakthroughs in the futuristic ones will be necessary in order to make anything like a shield work. These technological breakthroughs do not mean inventing new types of lasers, but they do mean increasing the power of current lasers by factors of 100 and more. It has now become abundantly clear that many of the early hopes for the X-ray laser were grossly overblown and that high-tech weaponry like this has not lived up to expectations.[10] Weapons scientists charge that Star Wars project leaders and senior scientist Edward Teller actively misled political leaders in Washington. Everyone in the community agrees that futuristic weapons like X-ray lasers are much further from reality than was thought when Star Wars began.[11] The only argument in the community is whether the overly optimistic statements were simply an error of judgment or whether they were part of a deliberate effort to mislead people and obtain additional support for the Star Wars program. One critic dubbed these efforts as "hawking nuclear snake oil."[12] Because the futuristic technologies are so far away, and because SDI management has sought to deploy a system in the 1990s, traditional technologies like "kinetic kill vehicles" which destroy incoming missiles by physically hitting them have received more attention.

In the last five years, scientists have developed a growing appreciation of the difficulties of controlling any massive defensive system. No manual system could handle 10,000 missiles at once, and the SDI system would have to be controlled by computers. The rapid pace of a battle in which 10,000 missiles had to be identified, tracked, aimed at, and shot down in a few minutes (assuming that the Soviet Union throws a sizable part of its nuclear arsenal at us at one instant) would make the final battle scene in the movie *Star Wars* seem like slow motion. These computers would rapidly and reliably have to distinguish enemy missiles from civilian aircraft and from decoys. While the computer systems could be tested in simulated battle situations, no one would know for sure whether they would work in real battles unless Armageddon struck.

As a result of five years of work and debate, SDI discussions by technologically knowledgable people are no longer about impenetrable shields, although newspaper headline writers continue to use the word "shield" in referring to the program. No one associated with the SDI program has claimed 99% efficiency, and as early as 1985 SDI Chief Scientist Gerald Yonas and Lieutenant General James Abrahamson, head of the SDI program, were explicitly denying that a shield is the goal.[13] Indeed, the system which SDI advocates in the Pentagon would like to deploy in the 1990s is far from a shield. This system, costing $50–100 billion, is advertised as being able to stop only one third of the incoming Soviet missiles.[14]

Three recent events have highlighted the difficulty of keeping out all incoming missiles in realistic situations. In the late summer of 1984, Soviet aircraft were unable to distinguish between a Korean Airlines Boeing 747 which had strayed off course and a military surveillance aircraft.

A few years later, a German teenager named Mathias Rust was able to fly a small Cessna through what must be the most heavily defended part of Soviet airspace, near Leningrad, and land this tiny plane on Red Square in Moscow. Lest Americans think that only the Russians can make stupid mistakes, in 1988 the sophisticated Aegis electronic equipment aboard the naval vessel *U.S.S. Vincennes* was unable to tell the difference between an Iranian F-15 fighter airplane and a civilian A300 Airbus and the civilian plane was shot down.

In all three of these cases, the defenders were facing the same type of situation which a "Star Wars" defensive scheme would face. In all three cases the defenders, whether human, electronic, or both working together, were not up to the task. The targets in all cases were relatively slow-moving aircraft, not missiles which can zip around the world in less than an hour. At least in the two cases where the Soviets were the defenders, there was plenty of time to make a decision on what to do. These cases show that SDI Chief Scientist Yonas is right in that 100% leakproof defenses don't exist.

There is the additional problem, raised 15 years ago in connection with a more modest defensive scheme, that the other side can use countermeasures to confuse SDI or to get around it. The original SDI scheme placed great weight on using lasers to intercept enemy missiles in the "boost phase" when their rockets were burning, when satellite sensors could easily detect the heat of rocket exhaust, and well before any decoys could be deployed. The other side could simply use bigger rockets and shorten the duration of the boost phase to make it much more difficult for SDI to detect the rocket exhaust, make sure that it was a military missile rather than a civilian mission, point a weapon at it, zap the missile, and verify that the target was indeed hit in the short time that was available.

Political support for the SDI program began to wane in the 1980s. Many congressmen began to realize that if the SDI program were ever to lead to a shield, it would be in the very distant future. Increasing concern with the Federal deficit meant that SDI money was coming from other Defense Department weapons programs. Lieutenant General Abrahamson, who had led the SDI program virtually since the President's 1983 speech, left the SDI organization, and the money spent on SDI no longer increased with each year as it did in the mid-1980s.

Democratic Senator Sam Nunn has made a very sensible proposal which puts SDI in perspective. It's quite clear that an antimissile system can shoot down one missile, and it's quite clear that it can't shoot down 10,000. The real limitations of an SDI program fall somewhere between those two extremes. Depending on how many missiles an SDI program can cope with, it might well make sense to deploy SDI at some time in the future when the nuclear arsenals of both sides have been significantly reduced and a major concern is protecting against accidental missile launches. For these reasons, Nunn supports continued research on the SDI concept at a relatively modest level, with the ultimate goal being an inexpensive system which could protect us against accidental launches. My own position on SDI is similar to Nunn's, except that I can visualize some other situations in which a modest SDI system might be useful. A world in which many nations had a few dozen warheads might well be a world in which a modest SDI system could protect us against an attack from, say, South Africa. It's even possible that such an SDI system would be deployed as part of some international consensus.[15] For these reasons, I personally support SDI research at an appropriate level but oppose immediate SDI deployment.

Antisatellite Weapons

Another Reagan Administration initiative, which attracted considerably less publicity but which may have considerably greater implications, was the resumption of American testing of antisatellite (ASAT) weapons systems. William Burrows refers to ASATs as the "netherworld" of satellite operations. In Air Force jargon, an ASAT weapon "negates" an enemy satellite by firing pellets at it, jamming its electronics, spraying paint on its camera lenses, or detonating a nuclear bomb near it to zap its delicate circuitry.[16] Because most satellites are built of lightweight materials, they are quite easy to destroy. The Soviet Union tested one relatively simple type of ASAT in the late 1960s. Satellites launched from the Soviet Tyuratam space center were chased by ASATs launched from the same place, on the same orbit, which homed in on the target satellite during the course of several satellite orbits. Twenty tests were conducted in 14 years, ending in 1982. About half were successful.

Americans remained inactive in the ASAT field for decades, but resumed work on a system during the Reagan Administration. The American system is considerably more sophisticated than the Soviet one, since the ASAT and the target need not be in similar orbits. The ASAT is launched from an aircraft, quickly ascends to the satellite passing overhead, homes in on the target using infrared telescopes to sense the target's heat, and bashes into it to destroy it. In 1985, the Air Force chose an American military satellite which still had two functioning scientific instruments on board for the first ASAT test. The ASAT worked, smashing a solar observatory into a hundred pieces which had discovered several comets in the process of hitting the sun.[17] Yes, our ASATs work.

Different analysts see different effects on the arms race from space weapons like ASATs or Star Wars. One group of analysts, generally in the United States military, believe that instability and uncertainty in the arms race are beneficial to the United States. Since no one sees Star Wars as a shield, the avowed purpose of deploying a system which won't protect our population is to increase the uncertainty that the other side has. A similar rationale is given for ASATs. The general philosophy is rooted in centuries of experience with warfare before the nuclear age, where a good strategy was to keep the other side off balance. Hundreds of years ago it was done with smoke screens.

Another group of political scientists, generally identified with the "arms control community," argues that uncertainty in the nuclear age is not always good. These analysts argue that the escalation of the arms race to space will greatly increase the amount of distrust on both sides, creating a potential instability in the arms race and still more escalation.[18] In such a climate it would certainly be considerably more difficult to carry out international cooperative ventures, possibly even civilian ones.

Deciding between those two positions is as much a matter of general political philosophy as it is a matter of technical expertise. While arguments over issues like ASATs often have technical overtones, deep down they always come to whether an individual believes that negotiation or confrontation is a better approach to dealing with high-tech weaponry. To me, negotiation is the better way to go. The advent of nuclear weapons has indeed changed the way that hostile countries must deal with each other, because for the forseeable future each of the superpowers will have the capability to destroy the other even when it is militarily or economically crippled; a superpower can still destroy the other side as long as it still has even a small percentage of its nuclear arsenal intact. However, I recognize that good arguments can be made for the other position.

Implications for the Space Program

The passive and active, simple and complex pieces of military machinery described in the last several pages share one common property: They are all machines. Spy satellites, ASATs, and even Star Wars can be accomplished with only machines in orbit. People are not required to live in orbit to support them. As a result, this type of military activity cannot be used to justify space settlement, even in low earth orbit. It will have little effect on deciding whether our future in space includes space settlements or purely scientific research, unless of course a space arms race leads to Armageddon and there is no future for humanity in space or on the ground.

Nevertheless, there is a close relation between the military and civilian space programs because both share the same technological infrastructure. Both programs need reliable launch vehicles to put hardware into orbit. Both programs need communications systems to maintain contact with satellites. Until recently, this system was a global network of ground stations. Starting in the late 1980s, the communications system, which will occasionally be used by both civilian and military satellite controllers, consists of several Tracking and Data Relay Satellites (TDRS, pronounced "Tee-Driss") in geosynchronous orbit. A TDRS was on board the ill-fated *Challenger* when it exploded in January 1986, and a replacement TDRS was on board *Discovery* when its launch in September 1988 marked the resumption of the American space shuttle program.

Another example of shared technology is the close relation between the spy satellites and particular astronomical telescopes. The Hubble Space Telescope (HST), a 90-inch telescope which NASA will launch into orbit in 1990 or so, will probe the universe to depths and in ways which are impossible from the ground. The mirror of the HST is about the same size as the mirrors used in the spy satellite cameras and is made by the same firm (Perkin-Elmer). No one can be sure, but it's quite possible that the mirror would simply be unobtainable were it not for Perkin-Elmer's expertise in producing similar mirrors for the military. Technology transfer can work in the other direction, too; astronomers pioneered in the development of "rubber mirrors" which could change their shape in response to subtle changes in the atmosphere. It is quite possible that the latest generation of spy satellites, the KH-11, uses the same technology.[19]

The examples discussed above illustrate the synergism between military and civilian space programs. Sharing technology and infrastructure can present problems when there isn't enough of them to go around. In the late 1970s and 1980s, the military and civilian space programs competed for the same set of resources and had an adverse impact on each other. From the military's perspective, the decision to throw away the giant Saturn 5 rocket and rely on the space shuttle as the only way to launch substantial payloads into space was clearly a mistake. It meant that the Air Force had no way to launch heavy payloads into space in the late 1970s, when the shuttle was behind schedule, and in the late 1980s when the shuttle was being fixed. As late as 1984, NASA was still trying to monopolize access to space; the Air Force had to overcome vehement NASA objections in order to purchase some Titan launch vehicles.

From the civilian perspective, competition for launch facilities also creates some problems. During the development of the space shuttle program, military requirements constrained the shuttle design, increasing its cost.[20] Once the shuttle started sending payloads to orbit in the 1980s the situation became even worse. Military and civilian missions had to stand in the same waiting line in order to take off. After *Challenger*, the waiting line became very long. As an extreme example, consider the Galileo mission to Jupiter, in which the scientific instruments were selected in 1974 and it will not arrive at Jupiter until 20 years later, in the mid-1990s. Virtually everyone in the space science community has a favorite set of missions which are sitting on the ground. I'm just waiting for the flights of the Extreme Ultraviolet Explorer (EUVE), originally scheduled for 1987, and now flying in the 1990s; the Astro space shuttle payload, with three flights scheduled for 1986–87 being reduced to one flight in 1990, and the HST, repeatedly delayed from 1984 to 1989.

SOLDIERS IN SPACE?

The military role in space depicted so far will have a minimal effect on the long-range future of the civilian space program, on affecting whether we settle the inner solar system, pursue a largely commercial space program, or devote the program to space science alone. Is it conceivable that future military activities could change this picture, creating a military reason to put soldiers in space?

In the 1960s, there were indications of an interest in a manned military presence in low earth orbit. The United States Air Force spent $1.6 billion seriously studying a manned military space station. This Manned Orbiting Laboratory (MOL) would be a small cylindrical space station which would attach to a Gemini space capsule. They envisioned the whole MOL to be about the size of a small trailer, 20 feet long and 10 feet in diameter. The story of the MOL program is all too familiar; estimated costs escalated through the 1960s, the target date of 1967 came around and actual space hardware was no closer at hand than it was when the program begun, and in 1969 the program was canceled.[21]

The importance of the MOL program to our present story is that even after five to six years of real work no real mission for the MOL emerged.

This project would not have come to a dead stop if there had been something for the military personnel in the MOL to do. But satellite inspection, reconnaissance, testing, interception of electronic signals—all of these tasks could be done better at the end of the decade by automatic, remotely operated machinery rather than by orbiting military personnel.

Since then, the Defense Department has shown no really strong interest in the manned side of the space program. While many space shuttle missions are military, virtually all of them have simply used the shuttle as a way of delivering cargo into low earth orbit. A few shuttle missions have included tests of military equipment; for example, a classified infrared sensor was on board and tested during the fourth flight of the shuttle *Columbia* in July 1982.[22] However, it seems that these activities are sufficiently modest in scope that they could not, by themselves, justify a manned space program, even in low orbit.

In principle, soldiers in space could repair and refurbish existing spy satellite systems. However, some thought about the costs of current hardware suggest that such a facility could not justify its existence, and that it would be cheaper to simply replace a spy satellite rather than repair it. Averaging the total costs of the space shuttle program over 20 years and an estimated 125 shuttle launches works out to a cost per space shuttle launch of a half a billion dollars.[23] If it took even one space shuttle launch to repair even one of the expensive, $300 million spy satellites, it would be cheaper to simply launch another spy satellite with an expendable rocket. While the economics of satellite repair might change in the future, the trend in high technology equipment in the last several years has been that repair costs go up and replacement costs go down.

BEYOND LOW EARTH ORBIT

With all that we can do in near-earth space, there is no conceivable reason to put military hardware beyond low orbit if the purpose of doing so is to support military units that are fighting on the earth. The moon is so much farther away. Why send a missile from the moon to the surface of the earth, where the trip will take three days, when the same missile can descend from low earth orbit in 15 minutes? Occasionally the moon has been used as a passive radio signal reflector, so that an orbiting satellite can communicate with a ground station without sending direct signals which the other side could intercept and use to infer the location of the satellite or the ground station, but this use does not require any hardware on the moon.

Indeed, military use of the moon and of other celestial bodies has been explicitly ruled out by the small amount of outer space law that does exist.[24] Extraterrestrial law, like other bodies of law, is based primarily on a few international treaties and conventions. Its principles are also drawn from customary practice, terrestrial law, and scholarly legal articles. Political scientist Christopher Joyner and former astronaut (and United States Senator) Harrison Schmitt have distilled a number of general principles from the 1967 Outer Space Treaty (signed by 86 nations), the prime source of extraterrestrial law:

- Space is the province of all mankind.
- All states should be free to explore and use space, including the objects in it.
- Space is not subject to national territorial claims.
- Space shall be used for peaceful purposes.
- International law formulated on earth can extend to space and celestial bodies.[25]

The "peaceful purposes" principle is enunciated particularly strongly in the treaty. It specifically forbids the establishment of military bases or the placing of weapons of mass destruction (in other words, nuclear bombs) on celestial bodies like the moon. Thus the treaty, if its influence remains, rules out a military justification for the settlement of the inner solar system.

In the United States, both political parties and the military supported the 1967 treaty, and the Senate ratified it with a resounding 88-0 vote. However, it does not make space purely peaceful. Putting bombs in earth orbit violates the treaty, but putting military machinery in orbit is permitted and, as shown above, has been done by both sides. Indeed, historian Walter McDougall has described this treaty as "all show and little substance."[26] A cynic's view is that the only activities ruled out by the treaty are those which the major spacefaring nations couldn't do, or didn't want to do, at the time. It's a start, but it leaves several open questions.

For instance, current space law has no formal provisions for dealing with the exploitation of resources, should this occur in the future. A similar problem existed with the Antarctic Treaty, which was recently amended to permit explicitly nations to establish bases for resource exploitation. Should we get to the point where space resources will be used, a similar amendment to the Outer Space Treaty will no doubt be made. There will unquestionably be considerable wrangling about whether the major space nations or whether the world at large will reap the lion's share of the benefits from this exploitation.

With two exceptions, the treaty's statement that nations can't claim title to celestial bodies is simply a reflection of the large size of these objects and the vulnerability of military installations on them. The treaty was formulated when the only relevant celestial body was the moon. The moon, and bases on it, are basically indefensible. Because space installations are in a hostile environment, they are very difficult to defend against attack; were a lunar base to be "hardened" so that even a modest-sized bomb wouldn't disable it by destroying some critical piece of hardware, it would be buried so deep beneath the lunar surface as to be useless. The moon is so large that it would not be possible to prevent an adversary from landing somewhere on it.

There would not really be much point to attacking or defending a particular base on the moon, either. If an attacker's objective were to establish a human presence on the moon, there's plenty of lunar surface to work with; in this respect, the moon is like Antarctica. The only situation in which I can imagine war breaking out on the moon is one in which there are a few lunar locations which are unique enough to be worth fighting over. If water exists in only a few polar craters, those craters might turn out to be very valuable commercial assets, and an adversary might seek to take one of them over by force instead of establishing a base in the one next door.

However, there are two locations in the inner solar system which could both contain unique resources and which are small enough to be defensible in a military sense. Many people in the space community have wondered whether there will be any particularly strategic locations in the solar system which will generate conflicts in the way that strategic harbors or straits have done in the past. Some experts speculate the two Martian satellites, Phobos and Deimos, could be two such strategic locations. They are small enough to be defensible, and they could be quite important.

Let's consider a possible scenario. Suppose that, by the end of the 21st century, there is some sort of an outpost or even colony on the surface of Mars. Resupply trips back to the earth would then be taken with some degree of regularity. If water is available on Phobos or Deimos, it will undoubtedly be used for something, most likely to refuel and resupply a ship returning to earth.

Phobos and Deimos are not only worth defending; they are defensible. One reason that the moon is not militarily defensible is that it is so large. But Phobos and Deimos are only tens of miles across, small enough that a small number of battle stations located on their surfaces could prevent another nation's spacecraft from landing on them. It's an interesting thought.[27]

While this all sounds a little grim, we will have had a fair amount of experience with the Antarctic and with the sea bottom by the time we establish bases on Phobos. The sea bottom is generally agreed to belong to no one, yet it may contain some economically valuable nodules of manganese and related minerals. The economic potential of the Antarctic is beginning to emerge from its icy wastes. In both cases, nations are trying to negotiate reasonable arrangements with each other which will permit operations like mining to take place without military battles. The Outer Space Treaty is too vague to handle a possible shootout at Phobos, but our other experience may enable nations to avoid such a 21st-century confrontation.

* * *

Both American and Soviet armed forces use near-earth space extensively and passively, placing thousands of satellites in low earth orbit, geostationary orbit, and elliptical orbits in between. These passive systems will have little effect on the future of humans in space, since they all orbit near the earth and do not require the presence of human beings.

Even the possible placement of active weapons systems, ASATs and antimissile systems, will not profoundly affect our future in space, except to the extent that both end up competing for the same scarce shuttle launch slots and both military and civilian space programs become crippled by the delays. Even these do not require humans in space. Indeed, the single attempt in the United States to justify a manned facility in low orbit, the Manned Orbiting Laboratory, died for want of a mission.

Possible military activity beyond low earth orbit is a bit more speculative. Most celestial bodies are not militarily defensible, nor are they sufficiently unique to be worth defending. Two possible exceptions are Phobos and Deimos, which are small enough to be defensible and may be in a key position for the future of solar system exploration.

CHAPTER 15

Exploration of the Near and Distant Universe

Space science, the use and exploration of space to advance human knowledge, has always existed in an uneasy symbiosis with the national space programs in both the United States and the Soviet Union. Space scientists and policy analysts agree that much of our activity in space, particularly in sending astronauts to destinations like the moon or space stations, has been undertaken for reasons that are basically nonscientific, to enhance national prestige and power. However, once astronauts are on the lunar surface or in a space station, those who sent them there for other reasons realized they needed something to do, and scientists realized that they had a tremendous opportunity to explore the universe in ways that would not be possible otherwise.

Scientists rapidly discovered that there is a great deal to be learned about our own planet, other planets in the solar system, and about the distant universe by working in space. Initially, astronauts had a role as lab assistants, performing experiments, picking up rocks, and placing geophysical instruments on the lunar surface. Later on, automated instruments played an increasing role in space science, leading to disputes between space scientists and space enthusiasts about the role of astronauts in the space program.

Because space science covers such a broad area, it is impossible to treat it all in a short chapter, or even in a book. As a result, in this chapter I will focus on two separate space science questions which can give a flavor of what we can learn from exploration of the near and distant universe. First, how does the greenhouse effect warm our planet's surface, and can we keep human production of carbon dioxide from warming our planet's surface too much? Second, do planetary systems exist around other stars?

Both of these promise to be hot topics for the remainder of this century and perhaps for much of the next.

THE GREENHOUSE EFFECT

For a number of years, atmospheric scientists have suspected that the earth as a whole is getting warmer. The hot summer of 1988, combined with the relative maturity of our understanding of the way the atmosphere traps heat on the earth, generated a renewed interest in the greenhouse effect and in various apocalyptic predictions of global warming, which would result in melting the polar ice cap, and flooding most of Florida, Manhattan, and other places.

The greenhouse effect keeps the surface of the earth warmer than it would be without an atmosphere. Visible light comes in from the sun and strikes the surface of the earth. On average, about one third of the sunlight bounces off the surface of the earth and clouds, and is reflected back into space. The remaining two thirds is absorbed by the earth's surface and contributes to making the earth's surface warm.

If the earth had no atmosphere, this would be the full story. But the earth does have an atmosphere. The earth's surface warms up and emits infrared radiation simply because it is hot. The atmosphere absorbs part of this infrared radiation and re-emits it to outer space and to the surface of the earth. Because the infrared radiation is trapped between the atmosphere and the ground, the surface of the earth absorbs not just the visible light from the sun but also the infrared radiation from the atmosphere. The more that infrared radiation is trapped, the more it gets reradiated to the surface of the earth, and the warmer the earth gets. The result is a planet which is then warmer than it would be in the absence of an atmosphere.

For the several billion years of recorded geological history, a delicate balance which we don't thoroughly understand has kept the temperature of the surface of the earth in the relatively narrow range where water is liquid but where the earth is cool enough so that living plants and animals don't overcook and die. Thanks to this remarkable balance, life was able to get started on the earth and continue to exist for the 4.5 billion years of geological time that have passed since the earth and the sun were born.

The concern of many scientists today comes from the role that human activities play in increasing the amount of carbon dioxide in the terrestrial atmosphere. Carbon dioxide is the natural by-product of most of the ways that we generate energy. Any time hydrocarbons are burned, the natural products are carbon dioxide and water. Carbon dioxide is not just a minor by-product of the reaction; whenever a carbon-containing molecule is burned in a chemical reaction, carbon dioxide is produced. The tailpipes of millions of automobiles emit carbon dioxide; pollution controls cannot get rid of it since carbon dioxide is the major end product of combustion. (Catalytic converters eliminate gases like carbon monoxide, oxides of nitrogen, and various hydrocarbons, which are all produced in smaller quantities.) Smokestacks of electric power plants emit carbon dioxide. Chimneys of individual houses, carrying the hot gases produced by the furnaces which keep those houses warm, emit carbon dioxide.

In principle, an increase in the amount of carbon dioxide in the atmosphere can increase the degree to which infrared radiation is trapped and results in increased surface heat. Atmospheric scientists have hotly debated whether the effect of carbon dioxide generated by human industries is significant and, if so, what the effect will be. In the most apocalyptic scenarios, the icecaps in the Antarctic and on the surface of Greenland melt and the sea level rises a hundred feet or so.

While human life would still go on, the economic consequences would be devastating. Florida, the seventh most populous state in the United States, would be almost completely inundated. The country of Bangladesh would simply cease to exist. Most port cities like New York, built at the ocean's edge, would go underwater.

Our need to understand the greenhouse effect better is underscored by the difficulty in finding any easy way to reduce the amount of carbon dioxide which is a natural, almost inevitable by-product of industrial civilization. The only ways to generate energy without producing carbon dioxide are to use solar energy directly or indirectly, or to use nuclear energy. Hot water heaters on a roof or solar panels which absorb sunlight and generate electricity use solar energy directly. This direct use of the sun's energy plays a very minor role in the energy budgets of most world economies in the late 1980s. Water falling over a dam, generating electricity, uses the sun's energy indirectly; the sun's energy, by driving the global water cycle, is responsible for putting a substantial amount of water high in the mountains.

The space program has two roles to play in improving our understanding of how the global weather machine works and whether human activities are warming our planet because of the greenhouse effect. First, cameras and other instruments located at the orbital vantage point can obtain a global perspective of our atmosphere which cannot be obtained in any other way. Measurements of the content of various gases which are responsible for creating the greenhouse effect have to be made over the entire globe if we are to understand their pervasive effects on the entire atmosphere. Measurements from a handful of ground stations, which are unevenly distributed over the surface of the earth, don't help very much.

But an even better understanding of the greenhouse effect is likely to come from an understanding of what planetary scientists call the "Goldilocks" problem. The greenhouse effect on Venus, the planet next closest to the sun, is ultraintense, warming that planet's surface to the roasting temperature of 730 K or 850 degrees Fahrenheit. The surface of Venus is, apart from the surface of the sun, the hottest place in the solar system. Mars, as we have seen, is quite cold. Just like the bears' bowls of porridge in the Goldilocks folk tale, Venus is too hot, Mars is too cold, and the earth is just right. Why? How did they get that way? And how could they stay that way for billions of years?

The answers to these questions will not come easily. We need to understand more about the ages and distribution of the Martian channels, the role of the polar icecaps on Mars in the history of water and carbon dioxide on this planet, and on the history of water on the planet Venus. In addition, we need to take a good, close look at our own planet from the orbital vantage point. We can make significant progress on all of these studies before the 21st century is too old.

THE SEARCH FOR EXTRASOLAR PLANETS AND BEINGS

Are we alone? This is one of the deepest questions human beings have been asking ever since we first understood that the stars are indeed other suns, and perhaps before.

Planets

We can ask this question in a couple of different ways, both of which are rapidly being answered by activities closely connected with the space program. The astronomers' first task is determining whether there are planets around other suns. Observations from space, combined with complementary work with telescopes on the ground, are just beginning to provide some definitive answers. A second multi-disciplinary effort on NASA's part, its exobiology program, is producing some startling new insights into the question of how life got started at all, and on different ways to search for life elsewhere.

Finding planets around other stars is a tricky business. Planets only shine by reflected light, emitting a feeble glow which is easily overwhelmed by light from the star itself. The only possible way to detect this reflected light is to launch a telescope above the earth's atmosphere, where the images are much, much sharper, free from atmospheric blur. Quite a few optimists plan to use the Hubble Space Telescope (HST), a 7.5-foot-diameter telescope scheduled for launch in early 1990, to search for planets around nearby stars, despite predictions by some that the instrumentation can't quite see these objects.

If the Hubble Space Telescope is not powerful enough to see planets around other stars, bigger telescopes can be built and deployed in earth orbit, and one is planned for the 21st century. The size of a space telescope is determined by the size of the largest-diameter tube which the space shuttle can carry up into orbit, and also I suspect by the size of the rather similar mirrors being made for spy satellites. A telescope much larger than the HST could produce far sharper pictures, making it easier to sort the feeble planetary light from its parent star.

Indirect Evidence for Planetary Systems

A number of indirect techniques have given us some tantalizing suggestions that planets the size of Jupiter, or slightly larger objects called "brown dwarfs," might exist around a number of nearby stars. One such way of detecting the existence of a Jupiter-sized object around a nearby star is to look for evidence of its effect on the motion of the nearby star. Strictly speaking, Jupiter does not orbit around the sun; rather, Jupiter and the sun orbit around their common center of mass. Because Jupiter is there, the sun does not travel through space along a perfectly straight line. Two independent groups of scientists, using telescopes based on the ground, have claimed to find Jupiter-sized objects around a number of nearby stars. The solar system is not unique.

We have made more progress on efforts to find indirect evidence for the existence of solar systems around nearby stars. Clouds of dust around a nearby star would absorb some light from the star, heat up, and re-emit this light as infrared radiation. The first space telescope which would detect infrared radiation, the Infrared Astronomy Satellite (better known by its acronym IRAS), was launched in the early 1980s. A surprising finding of IRAS was that many nearby stars were surrounded by dust clouds that emitted infrared radiation. The infrared brightness of a number of stars, including the bright star Vega which passes overhead in the afterdinner sky of early August (for people in the temperate zone of the northern hemisphere), was more than it should be. The only enduring interpretation of this infrared excess depicted these stars as surrounded by a swarm of dusty particles, ranging in size from ants to baseballs and possibly even larger. In the case of one celebrated star, with the improbable name of Beta Pictoris, astronomers Bradford Smith and Richard J. Terrile were able to take a picture of the disk itself. Our current understanding of the formation of stars and planets postulates that dusty disks like the ones which IRAS found around a number of nearby stars are the precursors of planets, or perhaps the leftovers after most of the dust has been eaten up by the process of planetary formation.

Another discovery, more indirectly related to planets, is the recent discovery of objects intermediate in size between the largest planet we know of—Jupiter—and ordinary stars. By popular consensus astronomers have agreed to call these objects "brown dwarfs." The discovery of brown dwarfs has had a rather checkered history, with "discoveries" being announced, only to go away when future measurements failed to confirm the initial finding. However, in the late 1980s, Ben Zuckerman and Eric Becklin, using an infrared telescope in Hawaii, located a brown dwarf companion to a particular type of star called a "white dwarf." White dwarfs are the final resting phases of stars like our sun when they finish going through their life cycles. White dwarf stars should produce very small quantities of infrared radiation, and so if a brown dwarf were present as a companion to a white dwarf, it would be detectable as an infrared excess. One such brown dwarf was detected as a companion to a white dwarf.

Brown dwarf stars are not planets, and some scientists argue that they are more closely related to stars than to planets. However, they are smaller than stars, and the discovery of one type of substellar object, the brown dwarf, strongly suggests that other types of substellar objects, namely, planets, exist. The brown dwarf story also illustrates that when our ability to go into space results in a new view of our universe, in this case through infrared radiation, we often discover new types of objects in it.

Is There Anyone Out There?

The most tantalizing possibility in this area is that we may actually pick up evidence of extraterrestrial intelligent beings. Nearly 30 years ago, Giuseppe Cocconi and Philip Morrison called attention to the radio part of the electromagnetic spectrum as being the most logical way that extraterrestrial beings might communicate with each other, and with us.

Radio waves carry very little energy, and the cosmos is quite quiet in the radio part of the spectrum so that there is very little competition.

Ever since 1961, radio astronomers have been engaged in searches for signals from extraterrestrial intelligence. Most often, these searches have been relatively sporadic, sometimes by-products of other efforts to use radio telescopes. In the last decade or so, these sporadic searches have produced clear indications of how contemporary computer technology can be put to use to make the search for extraterrestrial life both more systematic and more productive. As a result, NASA has embarked on a financially modest, but potentially very profound, effort to search for radio signals from extraterrestrial beings. Private individuals and organizations such as movie mogul Steven Spielberg and the Planetary Society have been invaluable in supporting this project.[1]

* * *

Space science, of course, includes a wide variety of types of investigation. I selected the greenhouse effect and search for other solar systems as examples to illustrate the extent to which space science can provide unique insights into profound problems. Space scientists can probe the origin of the universe, analyze celestial explosions, probe the global weather machine in unprecedented detail, compare different planets to each other, discover unique landscapes on the satellites of the major planets, analyze the diverse physical environments around the major planets, and explore the wide variety of celestial bodies which exist in the near and distant universe.

This chapter has focused on two relatively narrow areas of space science, in part to provide examples of what space science has done and will continue to do, given sufficient opportunities. Knowledge of the potential damage that the greenhouse effect can cause on the earth has created considerable concern in the late 1980s. While global measurements of the earth's atmosphere, only possible from space, can play a role in our increased understanding of the problem, it's also quite possible that comparing our planet to Venus and Mars can generate a better understanding of the way that the greenhouse effect works. A second space science venture, more far-reaching in scope, deals with the discovery of solar systems beyond our own. We have discovered "brown dwarfs" that are neither planets nor stars, falling in between the two, and dusty disks which may be related to planet formation. But planets themselves remain almost beyond our instrumental grasp, tantalizingly out of reach of our current observational capabilities.

CHAPTER 16

Future Space Scenarios

Now it's time to pull all of the disparate threads in this book together. What will happen in space in the next hundred years? No one can predict precisely what will happen, but it is possible to outline various alternatives, in a speculative vein. Predictions like those in the following pages are primarily made in order to expand human horizons and stimulate the thought processes rather than lay out detailed blueprints for the future. Such predictions are useful to provide visions of future accomplishments, to motivate projects with long-term payoffs, and to guide near-term space activities. I hope that all these elements are present in this chapter.

Two basic questions have framed this book:

- Will commercial activities in space, particularly those justifying human settlement in low earth orbit and beyond, succeed?
- What materials are required to support life in space, and can these materials be obtained in space or must we launch them at great expense from the earth's surface?

Neither question can be answered for certain now. Yet our future in space will depend on the answers to each one. In what follows I'll delineate the various possibilities. Of course even if I turn out to be one of the more successful prognosticators of the future, the actual events of the next hundred years will be some kind of blend of the scenarios depicted here. And various unpredictable events will inevitably pollute the picture, making some of these scenarios obsolete. I suspect that, 40 years from now, I will look over my reddish-gray beard (I hope there will be some red left in it when I'm 80!) and find some of these predictions on the mark and some, especially the dates, to be quite funny. Nevertheless, the nature of the space program combined with the joy of prognostication encourages a look at the long-range future.

Such prognostications are more than just mind-expanding; they can suggest some immediate actions which may be essential forerunners of a visionary space program in the next century. With a space program planned with an eye to the future, the positive answers to the two key questions posed here can be established (if they are indeed true) and the space settlement scenario that all enthusiasts dream of can be on the way to happening. The wrong kind of space program would focus exclusively on the megaproject of the moment, and these future questions would remain unanswered. In such a case, the ephemeral public enthusiasm for space exploration—which still exists even 20 years after the lunar landing—will eventually run its course, and Wall Street bankers will make the decisions about whether it makes sense to dream about mining gold among the near-earth asteroids. While Wall Street bankers can be very smart, they generally are not inclined to make investments in schemes which will not make profits for decades. In other words, if we leave commercial activity in space up to the private sector, it won't happen soon, and it may well never happen at all.

SPACE SETTLEMENT

Most visions of our future in space describe this scenario in which humans spread out into the inner solar system in the same way that our ancestors spread over the entire earth ten millennia ago. Indeed, President Reagan's Space Policy set in early 1988 stated that the three major goals for the American space program were:

- Establishing a long-range goal to expand human presence and activity beyond earth orbit into the solar system.
- Creating opportunities for United States commerce in space.
- Continuing our national commitment to a permanently manned space station.[1]

However, these words from a White House press release don't necessarily mean that actual hardware will come down the line. The goals quoted above are not even from the official United States Space Policy, for the policy itself is classified. Transforming such policies into actual programs is the hard art of Capitol Hill politics, and at the close of the Reagan Administration it was far from clear that any of the three goals enunciated above could be implemented in practice.

If space settlement is to occur, in my view it requires an affirmative answer to both of the currently open questions. Space settlement will clearly require the provision of major life-support materials, most particularly water, from the space environment itself. Two technological developments are necessary if this is to happen. First, the life-support system needed to sustain human beings in space must be closed considerably more than it is now. In other words, a CELSS (where the acronym means *Closed* Ecological Life Support System, not merely *controlled* . . .) must exist. Full, 100% closure is not required, but something close to that is. Water must be used and re-used, rather than flushed down the toilet and out of the spacecraft as it largely is now. Second, since the life support system will always require some replenishment, the necessities of life must be obtained from some celestial body. Because most of the weight of the material which humans require is water, it is likely that the primary need of future CELSSs will be water or oxygen, which can be derived from water. The moon, Mars, Mars's satellites, and the near-earth asteroids all seem to be likely possibilities, with the moon being perhaps the least likely.

The second requirement for space settlement demands that these colonies become self-supporting in a genuine commercial sense. Throughout all of human history, there have always been explorers willing to risk their lives to extend the frontier. In shorter supply have been the financial backers, be they taxpayers or kings, who are willing to stay at home and pay for the costs of the expedition with only slim hopes of immediate financial return. In fact, many human expeditions before the space age have realized tangible benefits, direct or indirect, financial or strategic. The most surprising such expedition was Magellan's, from which a haggard remnant of only 13 of the original crew returned with enough valuable cargo to make a profit.

It is popular in the space prognostication business for authors to come up with a sequence of likely events, along with dates; I will follow this tradition by outlining the next hundred years in space. The dates aren't intended to be meaningful in an absolute sense, since missions are often postponed and dates are not firm until just before launch. However, these speculative events will give you some idea of the sequence of events which must transpire, in my view, if the space settlement scenario is to become a reality.

Now Soviet Union invites foreign scientists to help in missions; United States shares tracking data.

1992 Laboratory tests with carbonaceous chondrite meteorites demonstrate the feasibility of asteroid mining.

1993 Ground tests demonstrate proof of concept of a CELSS, that a CELSS will work in principle.

1996 The first joint Soviet–American mission since the Apollo–Soyuz 1976 project, a Mars lander with a mobile rover, is launched.

1996 Soviet astronauts demonstrate a way to avoid bone thinning.

1997 United States space station is "in place" (whatever that means).

2000 German microgravity research conducted on board the Soviet Space Station Mir 3 uncovers several new and unexpected types of solid materials, including a new rare earth–metal–oxygen compound which is superconducting at room temperature.

2005 A first "fully" private microgravity laboratory module, paid for and operated by a Japanese consortium led by a major Japanese company, is attached to the American space station.

2005 A CELSS is successfully tested in space.

2010 ESA (the European Space Agency) lands a remote prospector on a near-earth asteroid and returns samples from it.

2015 The first privately financed space station is built.

2020 CELSS systems are routinely used in the space stations and modules of all four major spacefaring countries (the European Space Agency, Japan, the Soviet Union, and the United States).

2020 A pilot plant recovers water from the surface of the Martian satellites Phobos and Deimos.

2020 The first child is born in space.

2025 Total sales of space products, combined with sales of services (microgravity and ultrahigh vacuum laboratory use fees), exceed $1 billion for the first time.

2030 The first pilot plant for water and carbon extraction lands on a near-earth asteroid and begins remote operation.

2035 A cooperative expedition with Japanese, European, American, and Soviet participants lands on the surface of Mars.

2040 The Gross Space Product (total sales of all goods and services which depend largely on the existence of a space program) exceeds $20 billion for the first time.

2050 The population of people resident at any one time in low earth orbit exceeds 1000.

2060 Taxes contributed to world governments from space industries total $10 billion.

2080 Martian Resources, Inc., a global, privately financed corporation, using a 2% surtax on all space activities, builds a set of cycling cargo carriers which will return supply water and organic materials to LEO from the Mars system.

2085 Martian Resources, Inc., establishes a human outpost on Phobos in order to repair the cycling cargo carriers and to invent better ways of extracting more resources from Mars and its moonlets.

2090 A gold and platinum residue is successfully extracted from a near-earth asteroid for the first time.

2091 The Gold Rush to the near-earth asteroids begins.

A number of general characteristics of this future scenario are worth commenting on; while the dates aren't intended to be taken too seriously, the sequence of events is at least plausible. I've deliberately and not too subtly introduced a number of other characteristics.

Most important is that the key events in this timeline focus on technology, rather than on getting to some particular destination. The widely discussed landing on Mars is only a small part of the picture, and I have put it considerably later than it would occur if we (meaning the Soviet Union, the United States, or both) devoted all of our energies to this one single goal. In some ways, it could almost be left out of the timetable entirely. This mission serves more as a symbolic focus of our future efforts in space than as an end in itself.

The timeline above includes a variety of new participants in the space program, most importantly a number of hypothetical private companies. Space settlement simply will not happen if tax dollars are to foot the entire bill for space infrastructure. Private companies will have to finance a number of items, starting with individual space experiments as is happening now, working up to individual modules on space stations, and eventually, perhaps, financing a full space station.

None of this private activity will happen, of course, without some commercial return. For the sake of clarity, I have listed a particular product, discoverer, and discovery site in the timeline above: room-temperature superconductivity, discovered by German scientists, on the Soviet Mir space station in the year 2000. Superconductors are materials in which electricity can flow without resistance. The discovery in 1986 of superconductors which worked at liquid nitrogen temperatures (77 K) has revived the study of superconductors from the coma it had been in for a decade or more. If these materials have the right properties, they will have a large number of innovative commercial applications in such things as superfast computers, medical devices, and possibly in magnetically levitated trains, where magnets floating above the surface of superconductors would take the place of metal wheels on metal rails.

Why German scientists on board the Soviet Mir space station? Germany has probably been more active in microgravity research than any other Western European country; an entire spacelab mission was devoted to microgravity and paid for by the Germans at NASA's bargain basement price of $65 million. One could make a case that Germany has been more active than the United States. In addition, the German space community has been given opportunities so far denied to American scientists to participate actively in the Soviet space program. A particle detector, ostensibly from a German laboratory but in fact built by a group led by American space scientist John Simpson, was on the Soviet mission to Halley's Comet.

While there has been some limited interchange between the Soviet and American space programs, no American hardware has been linked to a Soviet mission since the joint Apollo–Soyuz mission in 1976, when an American and a Soviet space capsule docked in space. Sharing space hardware could be vital to the progress of the Soviet space program. The Soviets have consistently shown their inability to build space hardware that lasts; unless this changes, or unless some other country develops the ability to build reliable space machinery, America will continue to play a central role in space. However, at the moment, the United States has limited launch capabilities and little experience with long-duration space flight. My proposed date of 1995 for the Soviets' solving the bone-thinning problem may be too far off; in the late 1980s, they are sending astronaut after astronaut into space for long periods of time and they know far more than anyone else does about the hazards of long duration spaceflight.

The European consortium ESA (the European Space Agency), which has existed for decades, will also play a role. The sample return mission to an asteroid mentioned in the timeline as happening in the year 2010 is already in ESA's mission pipeline.[2] While ESA spends considerably less money than either NASA or the Soviet Union, it has identified a number of niches in the space program where it believes it can successfully compete.

This space settlement scenario remains confined primarily to low earth orbit; an outpost beyond occurs quite late in the game. I have postulated 1000 people inhabiting space stations in low earth orbit by the year 2050; if this happens, most of these people will be temporary residents rather than permanent citizens. If people are to live beyond low earth orbit for extended periods of time, there must be some reason for them to be there. Current and projected commercial activity in space primarily takes advantage of the unique properties of the space environment, which is just as good near the earth as it is on Phobos. The hypothetical Phobos settlement, listed as being established in 2085, will perform some of those few activities which offer the possibilities of some commercial return: spacecraft repair and supervision of resource extraction. Both of these activities may require humans on site, and cannot simply be done by robots on Phobos which are teleoperated by people on earth.

It's hard for me to see a human outpost on Phobos playing a larger role than the one which I have defined above, in spite of the fact that technology alone would allow it to be established earlier. While the whole time scale may be compressed, so that a settlement on Phobos happens within our children's lifetimes, I don't see such a settlement being moved up in the sequence unless it turns out that there may be some overwhelming commercial reason to send people out to mine Phobos, Deimos, or the surface of Mars itself. I have not seen any such proposals coming from the wonderfully imaginative community of space enthusiasts; perhaps we all should be thinking more about such proposals. Of course, until we know more about Phobos, Deimos, and Mars itself—and this information will come from robot precursor missions which are currently under way—it's hard to speculate meaningfully.

ROBOT MINES, FACTORIES, AND LABS

The space settlement scenario outlined above absolutely requires that the answer to the two crucial questions be positive—that we can extract life-supporting materials from the space environment and that we can produce some commercial return from placing human beings in space stations and on outposts in space and on celestial bodies. Other scenarios will emerge should there be other answers to these two questions. Suppose that the cost of supporting humans in space remains as exorbitantly high as it is now, but that space commercialization still can occur. What will happen then? How does the robot mines, factories, and laboratories scenario differ from the space settlement scenario?

Laboratories in low earth orbit need not necessarily have human beings living in them in order to operate. Advances in computer technologies, in communications, and in robot manipulative techniques, many fueled by the space program but some coming from manufacturing industries, have made it possible for human beings to operate equipment which is located at great distances. Several hundred astronomers, including myself, have used the International Ultraviolet Explorer (IUE), a telescope which orbits 22,000 miles above the Atlantic Ocean. We use this telescope by sitting in front of a computer located at NASA's facility in Greenbelt, Maryland. We type in commands to a computer, and the telescope moves. The telescope might just as well be on the roof of the building we are in as 22,000 miles away. In a similar way, astronauts on board the space shuttle can use the Canadian-built remote manipulator arm to construct space structures without the bother of a space walk. It is only a modest extension of current capabilities to imagine a human being on the ground, looking at pictures returned by TV cameras, and moving this remote manipulator arm around by sending it radio commands.

These two examples illustrate a number of ways in which human capabilities can be extended by having people and machines act in tandem, rather than independently. NASA has coined a term, "teleoperation," to refer to this type of interaction. Teleoperation is one of several ways in which people and machines can work together in order to accomplish various tasks; the various ways encompass varying degrees of human involvement and machine programming.[3]

Advocates of humans in space will correctly point out that by not putting people up there we are considerably limiting our future activities. Let's focus on a microgravity laboratory, as an example to explore the trade-offs. Such a lab, operated remotely, will not work anywhere near as efficiently as one with a person in it. Instruments will break or fail to work as anticipated and the operator will have to wait months for an astronaut to visit the lab and replace the balky component.

The remotely operated lab will have a very significant cost advantage for a private company that might want to operate it. That advantage doesn't exist today, but it will exist in the future. At the present time, companies that engage in joint endeavors with NASA do not pay the cost of sending their employees into space as astronauts. They go along with the whole space shuttle program, and frequently the shuttle flights that they go on have many purposes besides microgravity research. Such an astronaut is piggybacking on the whole NASA program which sends astronauts into space as part of a national effort to gain prestige, lead toward space settlement, and possibly land people on Mars. However, in the future, companies will have to pay the full costs of sending astronauts into space.

To roughly estimate costs, take an example of a hypothetical company, Space Vacuum Technologies, Inc. ("SVT"), which wants to establish a research lab in low earth orbit as part of the space station. SVT desires to have one astronaut present at all times to operate the lab, and the lab is to operate for ten years. A reasonable estimate is that SVT will have to pay $2 billion up front just to house the astronaut in the space station.[4] SVT would also have to pay something like $0.8 billion over the ten-year period to launch astronauts and relief crews and another $0.4 billion to cover its share of keeping the astronaut alive.[5]

There's more. The minute that a company wants to send some equipment into space along with astronauts, everything has to be what NASA calls "man-rated" (a sexist term that dates from way back when). The company must exhaustively test the equipment to make sure that it won't explode, emit toxic fumes, or do other nasty things in the space environment which will endanger the mission's safety. I can't even try to estimate the additional costs of all this testing, but experience with space science missions indicates that it can at least double the cost to equip a lab.

The total cost for ten years works out to $3.2 billion, or about $1 million per day. The cost will be high as long as the cost of maintaining people in space is high, as is called for in this vision of the future where CELSSs don't exist and all the life-support materials are ferried up from the earth's surface. A private company might understandably choose on economic grounds to lease part or all of an astronaut-tended space platform and accept the inconveniences of teleoperation. It has been just such a presumption which has led a number of private companies to design such space platforms. Messerschmitt–Bolkow–Blohm's SPAS (Space Pallet Satellite), ESA's EURECA (European Retrievable Carrier), and most recently Fairchild's LEASECRAFT are examples of such space platforms.[6]

This type of operation would not exclude astronauts from the space program, but their role would be more limited. Astronauts would go up in the space shuttle, replace some equipment in an orbiting laboratory and replenish supplies, and then come back down again. This "man-tended" mode of operating a space facility is considerably less expensive than operating a space station. (Again, the term is sexist, but it is the term NASA uses.) Indeed, this general concept of a space facility which would be leased for periods of time by private companies was proposed in 1988 by Space Industries, Inc., a private American company run by a number of former astronauts and other NASA leaders. The Industrial Space Facility (ISF) failed to attract a sufficient number of industrial customers, and its future for the short term is uncertain.

This particular idea may well be ahead of its time. Teleoperation and microgravity research are both quite quite new. The market for space goods and services is far from developed, as it must be if the scenario I'm discussing now is to happen. Furthermore, potential ISF customers probably have many opportunities within the NASA program. They can sign a Joint Endeavor Agreement (JEA) with NASA and let NASA pay for the costs of launching an astronaut into orbit. Smaller-scale experiments can go into space on board the shuttle as part of the Get Away Special (GAS) program, where a few thousand dollars can pay for a small slot in the shuttle's cargo bay.

A timeline for the robot space mines, factories, and labs scenario might look something like this:

1992 Lab tests on earth demonstrate feasibility of asteroid mining.

1997 American space station in place.

2000 German microgravity research conducted on board the Soviet Space Station Mir 3 uncovers several new and unexpected types of solid materials, including a new compound which is superconducting at room temperature.

2005 A proof-of-concept test of a teleoperated microgravity research laboratory is successfully conducted on the American space station.

2010 Pilot plant of a teleoperated space manufacturing facility begins operation in low earth orbit.

2015 A Japanese consortium orbits and operates a private microgravity research laboratory, and signs an agreement with NASA to send astronauts to tend it every year.

2020 American and Soviet space stations are abandoned as permanently inhabited facilities. Much of the hardware is transformed into power supplies and data relay stations for astronaut-tended space research, manufacturing, and space science facilities.

2060 Total sales of space goods and services exceed $1 billion.

2080 Lab tests on the Industrial Space Facility demonstrate feasibility of using the carbonyl process in a gravity-free environment to extract gold and platinum from asteroidal materials.

2090 A privately funded expedition to refine gold from a near-earth asteroid is launched.

Again, the speculative timeline is meant to be taken only half seriously. However, it does illustrate that abandonment of the regular use of astronauts in space is not the same thing as abandonment of space entirely. Some of the milestones of the space settlement scenario are left in. I even left asteroid mining in, after all. However, Mars takes a back seat now; a trip to Mars, which will probably happen for symbolic reasons is largely decoupled from the major focus of space activity: automatic machines at work in low earth orbit.

A SPACE PROGRAM LIMITED TO SPACE SCIENCE

Our future in space will take yet another form if the answers to the two key questions posed at the beginning of this chapter are both negative. In this scenario, extensive space industrialization will not happen. Commercial activities in space could well remain confined to the well-established communications satellite business and closely related activities, activities which could continue whether or not humans live in orbit.

The one significant space frontier which will remain is the use of space to expand human knowledge, or space science. The near Universe—the solar system—contains planets of a fascinating variety of shapes and sizes. Even if there is no practical application, surely the prospects of landing on the reddish sandy desert of Mars, of understanding the enigmatically complex millions of ringlets which make up the visible rings of Saturn, of understanding the answer to the cosmic "Goldilocks problem"—why Venus is too hot, Mars is too cold, and the earth is just right—appeal to many people. The bizarre satellites of Jupiter, Saturn, and Uranus contain some fascinating landscapes, and Saturn's giant satellite Titan may have oceans of liquified cooking gas. The more distant universe has its attractions too. The technology is at hand for discovering dust clouds and super-Jupiter-like brown dwarfs around nearby stars; discovery of Jupiter-sized planets is around the corner, and earth-sized planets are not impossible. Farther out into space, we have detected the enigmatic black holes, the shrunken remnants of burned-out stars whose super-strong gravity won't let anything, not even light, escape from them. Still farther out, at the edge of the universe, the space telescope will allow us to detect primeval galaxies, huge star-swarms which were formed shortly after the universe itself.

I would like to think that a nation's taxpayers could be persuaded that scientific exploration is worth supporting as an end in itself, not tied in with a multifaceted program like NASA's which pursues several goals at once. If so, then the 21st century in space will contain a number of space science highlights and new discoveries which cannot be predicted in some kind of timeline. All of the highlights of previous timelines, which focus on resource exploitation and applied research in space, would be absent.

Space scientists and space enthusiasts are in general agreement that a program devoted exclusively to space science would be very different from the space program NASA runs today. With the exception of two space science disciplines, astronauts would play little or no role. Most space science is done with automated equipment which occasionally needs repair. If repairs by humans are needed, such repairs can be done by individual shuttle missions. However, the cost of such a mission would be at least as high as the present cost of half a billion dollars, and it might not make sense to maintain a shuttle fleet just to fix instruments that don't work. It would probably be cheaper to simply replace them.

There are only two fields of space science where a human presence in orbit is absolutely required: microgravity science and the life sciences. Microgravity research has progressed from a focus on space manufacturing to a science discipline which studies the processes that take place in microgravity. If the science-only scenario outlined here is to take place, it will be phased in gradually, and several shuttle missions will have flown before we would have to make the decision whether these two fields alone could justify an inhabited or man-tended space station. So far at least, microgravity and space life sciences have had a relatively limited impact on the broader scientific fields which they fit into, and it would seem unlikely that the purely scientific aspect of these two fields could justify sending astronauts into orbit.

The success of any exploration program requires two ingredients: people who are willing to push back the frontiers, and the necessary support to allow those people to venture forth into the unknown. The changing character of exploration over the centuries, which historian William Goetzmann has described as a "First" and "Second" Great Age of Discovery, shows up in the contrast between Columbus the buccaneer and the scientists who have led the exploration of the Antarctic and who have played a significant role in the space program so far.[7] However, as I have argued earlier, the willingness of political authorities to spend public money on exploratory ventures is dependent on the probability of some kind of commercial or geopolitical payoff. In the absence of such a payoff, will it be possible to gain support for a purely scientific program of space exploration?

The European Space Agency's (ESA's) space program is occasionally cited as an example of a "pure science" space exploration program. However, ESA leaders frequently praise the program as one which can gain prestige for European science and thus for Europe in general, as a way of making Europeans feel co-equal with the American colossus. The space science "Horizon 2000" program is very specific on this point, describing ESA's program as giving "Europe the means of being an equal partner in a worldwide prospectus in space science, while honouring its cultural heritage and scientific tradition."[8] Furthermore, ESA's program is not just pure science. ESA and its member countries are heavily involved in sending their own astronauts into space, on board both American and Soviet spacecraft and in applied microgravity research.

If an ambitious, "science only" program of space exploration is to be supported in the same way that NASA is now, scientists will have to have much more success in gaining public support for the scientific sides of space exploration than we've had so far. I've had six years of experience in science public relations as the Education Officer of the American Astronomical Society. I found that it was quite easy to place stories in newspapers and interest politicians in supporting programs if they had one of three characteristics:

- Some connection with exploration by human beings. "No Buck Rogers, no bucks" is a widely quoted aphorism.
- Being the "first" in some well-defined way, such as being a possible entry into the *Guinness Book of World Records*. The first pictures of Mars, the first view of the Uranian rings, the most luminous warm galaxies in the universe, the first evidence for rings around Neptune, and the first discoveries of protoplanetary dust clouds—it was quite easy to convince the editors of newspapers or newsmagazines, such as *Time* or *Newsweek,* to cover these stories prominently.[9]
- Some obvious connection with technology and practical benefits. Erich Bloch, director of the National Science Foundation, was able to persuade Congress to commit itself in the summer of 1987 to doubling the Foundation's budget in a five-year period as a way of improving American technological competitiveness. This commitment was not met as a result of the stock market slump in October 1987.

Space exploration shares all of these characteristics, and as a result there has been considerable public support for it. However, without astronauts in the space program, it will be more difficult for scientists to sustain this level of public support, particularly once we have made the first visit to most of the interesting locales in the solar system. Follow-up missions where spacecraft orbit and land on a planet are often more exciting scientifically than the first missions where spacecraft simply zip by a planet and return a few dozen pictures. The public perception, in contrast, is that we are simply "returning" to a destination which we have been to already. There are a number of other pieces of information which indicate the difficulty of gaining support for a "science only" space program.[10] I'm personally convinced that space science will be a lot healthier if it accompanies a vigorous program of space industrialization, settlement, or both than if it has to stand on its own.

However, the scientific frontiers—which have only been lightly touched on in this book—still beckon, and many of them deal with deep human concerns about our origins and the origins of the universe we live in. Even after three decades of planetary exploration, we still don't understand just why our earth is so nice and habitable while neighboring Venus and Mars are so inhospitable. Where did the solar system come from in the first place? Are there planetary systems elsewhere? How did the universe evolve? It could well be that 21st-century scientists could obtain the support necessary to answer questions like these. It won't be easy, and the scope of a "science only" space program might not be much more extensive than the science part of the space program we have now. If there are limitations, though, they will come from the resources which the public is willing to allocate to space science, not from the lack of questions which scientists want to answer by sending equipment into space.

RESEARCH AND TOURISM

Let's suppose now that the responses to the two key questions result in a space program where it becomes cheap to support humans in space, perhaps by extracting resources from celestial bodies, but that space industrialization does not pan out. How will our space program differ from the science-only program just described? Microgravity scientists and life scientists would find it easier to include their activities as part of space science. In addition, if supporting humans in space became cheap enough, some wealthy people might even be able to become space tourists, paying the costs of a vacation in space.

I find the economics of this scenario to be a little difficult, but not impossible, to swallow. The primary way of making life support in space cheap is to make a CELSS work and to obtain water from some celestial body like Phobos. I wonder whether space tourism could generate a sufficient number of customers to justify a base on Phobos, whether tourism can generate enough human presence in space to make supporting life in space really inexpensive. Currently, the costs of keeping a person alive in space on a space station are about $1 million per day, as was worked out earlier. There probably aren't many people who can afford $7 million for a one-week vacation. However, if the costs could be cut tenfold, more customers appear; there are probably a few thousand people in the world who are rich enough to afford million-dollar vacations.[11]

One can imagine, then, what the Kennedy Space Center in Florida will look like in 100 years. Not quite the Kennedy Airport in New York, perhaps, but one can visualize 150 space tourists lining up at the two departure gates to take their seats on the two Orient Express hypersonic transports. The destinations listed above the check-in counter are Mir 23 and Mir 24, two of the 20 space stations which are set aside for tourist accommodations. In addition to the usual seat selection procedures, each of the passengers is given a standard physical exam, not unlike the standard physical given to payload specialists on the space shuttle a century before. PanSpace Carriers (a fictitious company, of course) just wants to be sure that it's unlikely that a space tourist will have a heart attack in orbit. The trip, while expensive, is quite short; it takes only ten minutes to reach orbit and another hour or so for the hypersonic transports to dock with the space station. The tourists then have a week to enjoy the pleasures of a gravity-free world, whatever they turn out to be.

AN EXERCISE IN PROGNOSTICATION

What will happen? It seems to me that the four scenarios which I've outlined—settlement, industrialization, science, and tourism— probably encompass the reasonable range of probabilities, given the presumption which I explained in the first chapter that the general characteristics of the global economy don't change catastrophically in the next hundred years. Which of these scenarios is most likely? It's amusing, and perhaps useful, to return to the two key questions, estimate the likelihood of positive or negative answers, and see which scenarios are most likely.

- Is it likely that we can obtain water from celestial bodies and significantly reduce the cost of supporting human life in space? My estimate is that the probability of a positive answer to this question is 70%.

- Will space industrialization happen? I think the chances are a bit better than 50–50 and will assume that the probability of a positive answer is 60%.

Estimating these probabilities is a bit like throwing dice, and in doing so I'm assuming that the answers to these questions are independent of each other. While this assumption may not be realistic, I have to make it in order to get anywhere. It then works out that the space settlement scenario, which requires a positive answer to both questions, has a 42% chance of happening. The robot mines, factories, and labs scenario, which has a negative answer to the first and a positive answer to the second question, has an 18% chance of happening.

To distinguish between the science-only and the tourism scenarios, an additional question comes up: Does tourism make any economic sense if it has to stand alone, not part of an overall space settlement scenario? I think it's a 50–50 proposition. In this case, the probability of a space-science-only scenario works out to be 26%, and the probability of a tourism scenario is 14%. (See Table 9 depicting the space future probabilities.) These numbers are rough guesses; if you like, you can make your own estimates and work out the numbers for yourself.[12]

* * *

This chapter has outlined a number of scenarios for the future of human endeavors in space. While it's unlikely that the future will work out exactly as my speculations predict, they probably encompass the range of reasonable probabilities. Space enthusiasts dream of space settlement, a future in which human beings establish colonies and bases throughout

Table 9. Space Futures

		Will Space Industrialization Work?	
		Yes	No
Can extraterrestrial resources be used to support humans in space?	Yes	Full space settlement (42%)	Research and tourism (14%)
	No	Robot mines, factories, and labs (18%)	Space science only (26%)

the inner solar system. Space could be industrialized without being populated, since robot mines, factories, and astronaut-tended research labs could do all the work. In a science-only scenario, most of the activity in space will be scientific research; if this happens, the space program will probably be no larger than it is today. The least likely scenario (in my view) is one in which some activity like tourism provides a justification for space settlement.

CHAPTER 17

The Trip to Mars

Prediction is always a gamble, but I feel quite confident that before the end of the 21st century, human beings will have set foot on the surface of the planet Mars. In fact, I rather suspect that the Mars trip will be complete well within the first half of the next century. Enough studies have been done so that we have a reasonably good idea of how the trip will be done and how long it will take. There are a number of new technologies which need to be developed if the trip is to come off; some are absolute requirements, and others can be worked around by considerably increasing the weight of the spacecraft. There are a number of open questions as to how the trip is to be undertaken. Will it be a Soviet expedition, an American expedition, or a cooperative expedition? How will this trip fit into the larger context of a space program?

A 20-YEAR MISSION, A TWO-YEAR TRIP

As we've seen, in some ways Mars is closer to the earth than the moon is. The amount of rocket fuel needed to make the trip to Mars is less than that needed to send a spacecraft of the same weight to the moon. Mars has an atmosphere which can be used to help a spacecraft orbit around Mars and land on the planet without using precious rocket fuel for braking. This atmosphere offsets the modestly heavier Martian gravity. While the distance to Mars, 35 million miles along a short but impossibly expensive trajectory, and nearly 200 million miles along the gentle sweep of the Hohmann transfer orbit, is great, most of the time is spent coasting.

However, the long distance to Mars can indeed prove to be a barrier to current technology. The Soviet Union has kept people alive in a space station for the nine months needed to get to Mars, but the Soyuz and Mir space stations are in earth orbit and can be resupplied from earth. Resupply, or repairs which require new spare parts, will be impossible for the team of astronauts who are making the Mars run. Furthermore, the nine-month trip to Mars is only part of the story; the astronauts need to survive the whole trip, enclosed in a small can, and never feeling the comforting tug of terrestrial gravity.

The Mission Scenario

Earlier chapters outlined some of the constraints involved in getting to Mars; the various loose threads will be tied together here with a Mars mission as a point of reference. There have been a number of studies, conducted by NASA and by private individuals working under the general leadership of an organization called the Mars Underground. The Mars Underground started as a group of students at the University of Colorado, and they have held a number of conferences in which Mars enthusiasts can pool ideas. As a result, we have some pretty good ideas of how the first trip to Mars will work, though there are still some decisions which will govern the final mission profile. The trip, as currently envisioned, will take two years from start to finish, and will incorporate a three-week stay on the Martian surface.

The total mass of the vehicles needed to go to Mars, land on its surface, and return to earth will be considerable; current mission scenarios call for a craft weighing from 1 to 3 million pounds. This craft is about 10 times as heavy as the Apollo craft which went to the moon, partly because of the larger crew size (currently visualized as seven or eight people) and partly because of the longer journey. The launch weight is uncertain largely because it's hard to forecast the amount of consumables needed to keep the crew alive for two years. The Mars craft will be launched in pieces, because no available launch vehicle can put that much stuff in low earth orbit in one trip.

Getting to Mars is not easy. The easiest path from one solar system destination to another is the gentle Hohmann transfer orbit, the elliptical path which curls between the orbits of earth and Mars. This path, which may become as well recognized as the great circle routes which are the shortest paths around the earth's surface, was worked out in the 1920s as being a natural way to get from one place to another. The difficulty with the Hohmann transfer orbit is that it is only possible to follow this path at infrequent intervals. It is possible to vary one's path from the Hohmann ellipse; modest variations require modest increases in the amount of fuel needed.

Part if not all of the hardware sent to Mars will follow this Hohmann transfer orbit on its outward path. Current "sprint" mission scenarios call for a cargo vehicle, containing fuel for the return trip, the descent vehicle, any surface rovers that the crew might need, and virtually anything else which is not needed on the outbound trip. This cargo vehicle would be launched on a pure Hohmann transfer trajectory first, and make a nine-month trip to Mars. The exact duration depends on exactly which one of the many launch windows is used.

The crew ship will be launched five and a half months after the cargo ship and follow a slightly shorter ellipse, taking 7.33 months to reach the red planet. The crew ship will not follow a Hohmann transfer orbit, because this mission plan has been developed to minimize the amount of time the crew spends away from home. However, because the crew ship doesn't include fuel for the return trip or any descent modules, only a modest amount of extra rocket fuel over that required by the Hohmann transfer orbit is required,

When both crew ship and cargo vehicle arrive at Mars, they will deploy a saucer-shaped aerobrake and dip into the thin layers of the upper Martian atmosphere. Air resistance will slow these vehicles down and prevent them from simply zipping past the red planet and heading off into the deep outer solar system. This will be one of the trickiest parts of the trip: Too much aerobraking and the crew ship will end up falling toward the Martian surface; too little, and the ship won't end up in orbit around Mars. Assuming that all goes well, the crew ship and cargo vehicle will match orbits around Mars and the crew will prepare for the descent.

At this point the crew of the Mars ship will split up in much the same way that the Apollo crews divided into a group of two people to land on the lunar surface and one person remaining in orbit. With a total crew of seven, three or four will remain in orbit, doing some science as well as transferring fuel and supplies from the cargo ship to the crew ship. The landing crew will use the descent vehicle, equipped with an aerobrake, parachutes, and rockets in order to land on the Martian surface. The total stay at Mars will be about three weeks, allowing the landing crew to roam about the surface in some kind of Mars rover.

The landing crew will have to take care to leave enough rocket fuel for the return trip to orbit around Mars. While the Martian atmosphere is a help in the descent, it is if anything a hindrance when the crew must blast off from the Martian surface, complete with bags of Martian samples.

They will then rendezvous with the orbiting crew ship, transfer themselves and the sample containers to it, and head back to earth.

If the landing crew were to wait until the next Hohmann transfer, until the minimum-energy launch window opened up, the mission would be very long indeed; a year and a half stay at Mars would be required. Such a mission would last for a very long time. Again, though, the sprint mission plan calls for a 165-day return trip. In this rather complex way the total amount of time people have to spend in space is reduced to a little more than 13 months. The total trip duration, counting the time that the cargo ship is on the way to Mars, is about 18 months.

Variations on a Theme

Soviet cosmonaut Yuri Romanenko's 11-month marathon stay on Mir, which ended in 1987, indicated that the zero-gravity environment is quite stressful physically and psychologically. As a result, the sprint mission plan went to great lengths to reduce the amount of time the human astronauts spent traveling to Mars. However, if we are to establish bases on Mars, the sprint approach can't be used for all missions, for it wastes too much rocket fuel. How else can one get to Mars?

The starting points for Mars mission planners are two basic classes of missions. A "conjunction" mission, which uses an absolute minimum of fuel, lasts for a total of three years, and the crew stays 1.5 years on Mars. Pure Hohmann transfer orbits are used going out and going back. In "opposition" missions, which are less complex versions of the "sprint" scenario, the total duration is about 1.5 years and the stay on Mars is much shorter, ranging from three weeks to a couple of months. The opposition missions use considerably more rocket fuel.[1]

A Venus swingby can modestly reduce the outbound trip time from the nine-month Hohmann transfer value and also obtain an additional opportunity to do inner solar system science.[2] The Venus swingby alters the spacecraft's trajectory in a maneuver called a "gravity assist." This trick slashes outbound travel time to about six months, at the expense of greater mission complexity. An additional difficulty is that the need for a Venus swingby tightens the constraints on exactly when the mission can be launched.

Another way of going to Mars, described briefly by the National Commission on Space, is to establish a group of "cycling spaceships" which would travel from earth orbit to Martian orbit and back.[3] While it's not likely that these ships will be in place before the first human expedition to Mars, it's certainly possible to get to Mars in this way. An advantage of this approach is that the symbolically important first human journey to Mars would create an infrastructure which would encourage future journeys.

The cycling spaceships envisioned by John Niehoff of Science Applications International, a consulting firm based on Chicago, would travel on elliptical paths which would take them past earth and Mars. If four of them were launched into orbit, someone seeking to go to Mars from the earth would wait until one passed by and then use a small spacecraft to match velocities with it. It would be like hopping on board a cosmic escalator. Once the cycling spaceship arrived at Mars 5.33 months later, the traveler would get on another spacecraft, hop off the cycler, and use aerobraking to slow down and go into orbit around Mars. The orbits of earth and Mars determine that it takes four cyclers to give space travelers a reasonably complete set of options.

A rather speculative variation on this theme is suggested by the possible presence of water on Phobos, Deimos, or the Martian surface. The great bulk of consumables needed for the return journey from Mars is water and its constituents (liquid hydrogen and liquid oxygen to make the rocket go and oxygen to replenish the air which leaks out of the spacecraft). If it is possible to establish a plant on Phobos or Deimos to remotely extract water from the surface of this object, such an outpost would be a logical predecessor to the first human mission to Mars. The initial cargo ship in this "strawman" scenario would then only have to carry those items which are not readily obtainable from Phobian or Deimian surface materials. ("Strawman" is a bit of NASA jargon; in the early stages of mission planning, a "strawman" scenario is set forth so that other people involved in planning can try to knock it down.) Since no mission planner now knows just what can be readily extracted from the surface of Phobos or Deimos, it's hard to say what more might be needed. A reasonable guess would be nitrogen (to replenish cabin air), some foodstuffs (spices? frozen meat? these would presumably be shipped from earth on a minimum-energy trajectory) to supplement the diet obtainable from the closed life-support system, and any consumables (such as liquid helium) which might be needed by some of the science experiments. The advantage of this approach would be a significant reduction in the mass which this initial cargo vehicle would need to carry.

Still another variation on the theme is suggested by the existence, on the drawing board to be sure, of a number of low-thrust, high-efficiency rocket technologies discussed earlier. These engines propel spacecraft with a series of gentle pushes, in the end requiring less fuel mass because the exhaust velocities are greater. The two current possibilities which provide cheap propulsion with low thrust are ion engines and solar sails. If these can be made to work, they would make a great deal of sense on most of the fuel-eating parts of the Mars trip. The velocity changes on the outbound trip can all be made gradually, for the sudden deceleration required when you reach Mars can be made via aerobraking. The only part of the Mars trip where substantial rocket power is absolutely required is the ascent from the Martian surface to orbit around Mars. Since the fuel for solar sails is free, the use of solar sails might make it possible to reduce trip times significantly.

NEW TECHNOLOGIES

If we can plan any number of missions which take humans to Mars, why aren't engineers developing definitive mission designs right now? Even for a mission which includes an absolute minimum of new technology, a number of new ways of doing things in space need to be developed. For years, NASA starved its basic aeronautics program to find money to allocate to other apparently more important programs like the shuttle. The space agency finally began looking toward the future, beyond the space station, by implementing Project Pathfinder in the 1980s. This project draws together and substantially supports a number of technologies which are either absolutely required or highly desirable in connection with a Mars mission. There are 18 separate projects which make up Project Pathfinder, and I will only describe a few of them and where they stand.

Aerobraking

Aerobraking is somewhat related to the aerocapture technology used by both the American and Soviet space programs in order to bring spacecraft back from earth orbit. American spacecraft of the Apollo era, and all Soviet spacecraft to the present day, used erodable heat shields on the forward end of the space capsule to help slow them down. Parachutes were then used to slow the spacecraft still further, to bring them to a landing in the ocean (for the Americans) or in Siberia (for the Soviet Union). The American Space Shuttle Orbiter uses ceramic tiles, which don't erode away, to bring the craft to a velocity where it can then fly like an airplane and land.

In aerobraking, a spacecraft would approach a planet with an atmosphere, following an orbit which dips into the atmosphere at high speeds. The spacecraft deploys a saucer-shaped shell to slow it down. Instead of plunging into a parachute-assisted landing on the surface, the craft reemerges from the planet's atmosphere, traveling slower. Repeated entries into the planet's atmosphere, combined with some assist from rockets, will eventually put the spacecraft into a circular orbit, above the planet's atmosphere.

The trick in aerobraking is to bring the spacecraft in at just the right angle so that it won't plunge into the Martian (or terrestrial) atmosphere on a descent trajectory and so that it won't simply skip off the atmosphere and head off into the great beyond. The spacecraft's pilot and computers must tune the spacecraft's orbit sufficiently accurately that it will go back out of the planetary atmosphere, yet slower—on the first pass, slow enough so that it will orbit the planet and dip into the atmosphere again. The concept which is currently being articulated is one where the incoming spacecraft uses the upper part of the planet's atmosphere to do most of the work, and it deploys a saucer-shaped shell in front of the spacecraft to slow it down. It all sounds nice and glib, but there are many details to be worked out: What is the shell to be made of? Will it be fixed in size? How can it be held in place stably? Answering these questions will require an extensive series of tests; the technology has not even reached the proof-of-concept stage yet.

Closed Life Support Systems

The weight of the spacecraft which will make the Mars trip is strongly dependent on the mass of water and oxygen which will be needed to support the crew. If the spacecraft is built in the Skylab or space shuttle mode, where the only thing that is recycled is the water produced by the fuel cells, the weight budget will be extravagant indeed. While the words Closed Ecological Life Support System (CELSS) imply a system which has no losses to outer space, it's highly unlikely that the losses will be reduced to zero. Nevertheless, any significant reduction in the amount of consumables which the astronauts need to carry, particularly in the water loop, can help reduce the weight of the spacecraft and the cargo vehicle significantly.

The new technology with the longest lead time is the possible use of plants or algae to regenerate the oxygen in the cabin air. The National Academy of Sciences Space Science Board, a blue-ribbon panel of space scientists, pointed out an obvious but so far unstated fact in a report released in 1988.[4] If astronauts, especially on a Mars trip, are to rely on plants to regenerate oxygen, the plants better stay alive. If all the plants die, so will the astronauts. Clearly, if astronauts are to rely on a CELSS for the Mars trip, it will have to be tested extensively. Because the trip lasts over two years, the tests will have to be at least that long. And if something goes wrong halfway to Mars, no rescue is possible—a great contrast to the situation in low earth orbit.

Water Supply at Mars

Some of the more speculative mission scenarios will require yet more groundwork. If any of the consumables which the crew is to use on the return trip are to be obtained from Phobos, Deimos, or Mars itself, clearly we need to be sure that the necessary materials can be extracted before we send the crew on its way. The Soviet 1989 missions to Phobos and Deimos will provide us with our first indication of just what the composition of these potentially very important objects is. But even surface composition measurements won't tell us how easy it is to extract sufficiently pure water easily, on an asteroid with virtually no gravity, where gravity-driven mechanisms for separating different chemicals from each other won't work.

There are a number of other technological problems which need to be solved before we can go to Mars. Establishment and use of fuel depots in space and transfer of ultracold, volatile liquid hydrogen and oxygen from one spacecraft to another in the absence of gravity are real challenges. With no gravity, liquids don't flow from high points to low points because there are no high points and low points. Liquids need to be pumped from one place to another. How much radiation shielding will the astronauts who are on Mars need, and how will they get it? What will provide the power to the Mars lander? Where on Mars is the best spot to land, both from the viewpoint of crew safety and from the viewpoint of scientific return?

With all of these unsolved technological problems, planning and designing the Mars trip will take at least one to two decades. Much of the necessary groundwork is interesting in its own right; the necessary exploration of Mars and its satellites will provide enormous scientific dividends even if, for some reason, humanity never sends representatives of our own species to the surface of the red planet. One of the key precursor missions, the American Mars Observer, was in fact suggested by the planetary science community completely independently of any consideration of future human missions to Mars.

WILL WE GO TO MARS?

As I write this, the planet Mars is a coppery red light high in the southern sky. I just went out to take a look at it, and I can't imagine that we would never go to Mars—not ever in the history of humanity—if we have the ability to do so.[5] The technology to make the trip is certainly in sight, if not in hand. Few people whom I talk to believe that we won't go to Mars—won't *ever* go to Mars—given that we can make the trip if we try. The surface of Mars has a symbolic importance which is at least equal to, if not greater than, the North and South Poles which were reached in the early part of this century. I simply cannot imagine that we would not go, ever, if we can go.

The reasons which have been expressed for making the journey are many, and the many individual supporters of the trip each have their own. Official declarations of support for the trip have emerged from influential editorial pages (e.g., *The New York Times*), from various congressmen, from the Planetary Society, a group of 100,000 space enthusiasts, and from individual astronomers. Hawaii's Senator Spark Matsunaga has written a book as a call to arms.[6]

My own reason for believing in Mars as a long-term goal rests on its role as a symbol of human achievement and on the ability of a Mars mission to act as a highly visible centerpiece in a broader effort to settle the inner solar system. In the event that such settlement makes sense—in the event that the "space settlement scenario" described in Chapter 16 comes to pass—it is still true that a great deal of infrastructure is needed in order to transform the dreams of self-supporting space settlements into reality. The high initial cost, long lead times, and substantial uncertainties associated with the initial investment in this infrastructure will probably mean that it would be difficult to justify the investment on the basis of numbers alone. The Mars mission could serve as a highly visible focus to help space enthusiasts persuade hard-headed bankers and budget-cutters that the investment was worthwhile.

What are the costs of the trip, considered by itself, neglecting other investments in ancillary infrastructure? Presuming that we can develop a life-support system, which is more or less closed, solve the bone-thinning problem, and solve the radiation problem without adding enormous amounts of mass to a spacecraft, a trip to Mars need not cost too much more than the Apollo trip to the moon. Remember that in terms of rocket fuel alone, the surface of Mars is easier to get to, though a bit harder to return from. NASA has commissioned a number of detailed studies which reach the same conclusions—of course using the same assumptions about the technology. Most estimates of the cost of a Mars trip put it at about twice the cost of Apollo, that's to say about $160 billion in 1988 dollars.

What does $160 billion dollars mean in terms of what society can afford? It sounds like a lot of money, but it really isn't so much. A realistic estimate of the per capita burden emerges from a calculation which assumes that the cost will be borne largely, if not exclusively, by the major spacefaring nations (the United States, the USSR, Japan, and Western Europe). A simple calculation shows that if these taxpayers alone shoulder the entire burden, the total cost is still only $160 per capita, spread over a ten-year period.[7] That's $16 per year per person to go to Mars. I think most of us would be prepared to pay the price of one dinner out per year to see it happen. I'm sure that some time in the next 120 years the necessary political consensus will arise.

Some alert readers may have noticed that I have not used the scientific knowledge that will come from the trip to Mars as a justification for the expedition. As a space scientist, I have made this omission deliberately. From science alone, there is no way to justify sending human beings to the surface of Mars on a cost-effective basis. While sending people to Mars will produce tremendously good science, we can operate scientific instruments remotely. Almost all of the science we would like to do—including a sample return mission—can be done without requiring a group of astronauts on the Martian surface.

WHO WILL GO?

From the present perspective, the Soviet Union and the United States are well ahead of the rest of the world in experience in sending humans into space and keeping them alive there. At the moment, there seem to be three possible groups of people who could go to Mars. The Soviet Union could do it alone. The United States could do it alone. Or the two major superpowers could undertake such a mission on a cooperative basis. A cooperative mission could either be a simply bilateral one, or could involve a number of other countries.

Clearly, at the present time, the Soviet Union is the single country in the best position to undertake a Mars mission by itself. The numbers game of who has spent the greatest number of man-days in space, and who can launch the greatest weight of hardware into orbit, indicators which are highly quoted but are often meaningless as an indication of a nation's space capability, are in fact quite relevant to the mission to Mars. Astronauts on Soviet missions have spent more than 5000 days in space as of 1987, while America's total is a mere 1800. It is the Soviet Union which has actually grown a plant, *Arabidopsis*, through its entire life cycle in space, from seed to seed.

In launch vehicle capability, the Soviets are also ahead now. Their ENERGIA rocket has the same capabilities as our old Saturn 5, the ability to put about 100,000 pounds into earth orbit. Though we had the Saturn 5, we shut down the assembly line and don't have it any more. While the space shuttle can put 60,000 pounds into earth orbit and bring substantial payloads back, the shuttle's unique ability to return large payloads to earth is not particularly relevant to the Mars trip.

An additional Soviet advantage which space scientists like myself are acutely aware of is their apparent ability to stick to long-range plans and long-range commitments, a strong contrast to the stop-and-go situation with NASA. Missions which are repeatedly postponed, threatened with cancellation, and reconfigured to fit new launch vehicles always end up costing far more than they would have otherwise. Developing the technology for the Mars trip will take decades; while the funding of Project Pathfinder is a promising first step, later stages of Project Pathfinder will end up costing considerably more. Will Americans be willing to make the necessary long-term investment? As an American, I certainly hope so, but as an observer of the fickle political process, I'm not so sure.

Despite the Soviet lead, the United States brings some strengths to a potential Mars mission which the Soviet Union lacks. While the Soviet Union has a technological capability in some areas of aerospace technology which is equal to America's, there is no question that the technological infrastructure in the United States is considerably broader. Computers will, no doubt, be quite important in spacecraft control during the aerobraking process, and computer technology is one area where the United States, though not necessarily preeminent in the world (Japanese competition is fierce), is clearly ahead of the Soviet Union. In addition, we have been able to build spacecraft which last, last, and last; the International Ultraviolet Explorer is a rather sophisticated spacecraft with an 18-inch telescope, and it has lasted for ten years, as has the Voyager spacecraft which has produced such spectacular images of the planets in the outer solar system.

A number of scientists in both major countries have suggested the logical alternative to a purely American or a purely Soviet mission: a cooperative mission. Indeed the strengths of the two countries' space programs are complementary; where the American program is weak, the Soviet program is strong, and vice versa. Specifically, we can use the Soviet experience with long-duration spaceflight to help design a better mission, and let the American aerospace industry build the hardware so that we can be more confident that it will last for the duration of the trip. In addition, a cooperative mission means that fewer resources will be required.

A number of voices, particularly in the United States, have been raised against a cooperative mission. Official NASA response has been guarded. I have participated in planning discussions of various cooperative missions between the United States and various European countries and have seen some of the hurdles which must be overcome in persuading people to work together. Developing the necessary sense of trust among people who have not previously worked together (and who may scarcely know each other) doesn't happen overnight.

Despite these problems, a cooperative mission has one distinct advantage: cutting another space race off at the pass. One reason that the Apollo mission left such a meager inheritance of important space hardware was that in our eagerness to beat the Russians to the lunar surface, and to meet President Kennedy's timetable of landing on the moon in the 1960s, some decisions with negative long-range consequences were made. When there was a choice between building a technology base for future space exploration and getting to the moon on time, speed always had the upper hand. In other words, how the trip to Mars is made could be as important as who makes it. A space race to Mars could lead to problems with the space program.

HOW NOT TO GO TO MARS: THE END OF THE SPACE PROGRAM

Many of us who have thought about the future of the space program for years find the prospects of a purely symbolic journey, especially a speedy, competitive one like the Apollo trip to the moon or Amundsen's expedition to the South Pole, to be quite troubling. A race to Mars could have disastrous consequences for the future of the space program in general. There was virtually unanimous agreement among the members of the

National Commission on Space and among the public who provided comments to this group that an Apollo-like, one-shot mission to Mars, which had to stand alone as a "space program" rather than fit within some broader institutional context, would be a mistake.[8] We clearly could make a symbolic journey to Mars without the benefit of a space infrastructure with long-term utility, but such a trip would leave us exactly where we were at the end of the Apollo program in 1970: unable to exploit this highly visible step beyond the confines of earth.

Seen differently and more apocalyptically, an Apollo-like jaunt to Mars could represent the end of the space program as we know it. The government-subsidized space programs supported by NASA in the United States and its bureaucratic equivalents elsewhere (IKI in the Soviet Union, ESA in Western Europe, and ISAS in Japan) could simply cease to exist, once someone's flag or flags fluttered in the red skies of Mars.

Let me start with the American perspective, which I am most familiar with, and recall the post-Apollo days. I believe the manned space program at NASA came perilously close to extinction in the early 1970s. NASA had not established any substantial post-Apollo plans involving human flight; many ideas were tossed around, but nothing had developed any substantial base of political support outside of the agency. Vietnam was consuming the nation's budget and its political will. Because of the way in which the Apollo mission was conducted, the only hardware which could be used in other space programs was the huge Saturn 5 rocket. There was not even the rudiments of a space station in low earth orbit which could serve as both a practical and symbolic stepping stone toward a more ambitious space program. We then of course made the ridiculous mistake of literally throwing the Saturn 5 away in the hopes that the space shuttle would be cheaper, and then focused on the shuttle as the central megaproject of the 1970s and early 80s.

To the extent that we understand the political decision-making process in the Soviet Union, there are very similar forces which could make a successful journey to Mars be the final climax of the Soviet space program. The government of the USSR is a tremendous bureaucracy in which appropriately placed individuals can gain access to tremendous resources. It's been often said that the Soviet Union is equal to or better than the United States in military power, in space, and in Olympic sports, but for the consumer it is a Third-World country. Without the inhibitions present in a democratic society, a bureaucratic government such as is found in the Soviet Union can allocate large amounts of resources to particular programs, either because they have tremendous symbolic value or because particularly energetic and talented individuals have been able to sell their programs within the bureaucracy.

In the case of the Soviet space program, both symbolism and strong personalities seem to play a role in developing the strong program that we see in the 1980s. The contrast between the crippled American space program and the Soviet program, with space stations which have been more or less continuously inhabited since the mid-1970s, has been on the cover of at least three major American newsweeklies in the late 1980s.[9] My conversations with Soviet space scientists have indicated their pride in their space program and their puzzlement about our mistakes. "Why doesn't your space shuttle work?" asked Soviet astronomer Geran Bisnovatyi-Kogan at a private dinner, held in connection with a scientific meeting in August 1988.[10] I could only vaguely mumble that the space shuttle was very complicated, feeling embarrassed that we had simply no way of launching heavy payloads into space. This conversation reinforces the clear impression from other sources that the Soviet process of allocating resources gives high priority to activities in which the Soviet Union can be demonstrably ahead of the United States in some dramatic field of human endeavor.

Astronomer Roald Sagdeev, head of NASA's Soviet Counterpart IKI ("Institute for Space Research" in Russian) in the 1980s, played an important role within the space program in encouraging more imaginative space science. For example, IKI under Sagdeev's leadership attached a Dutch/British X-ray telescope to the Soviet space station Mir, and this telescope was able to obtain critically important data on a very bright supernova which exploded in 1987. Even more important is Sagdeev's rise in the Soviet bureaucracy. He was elected to the Supreme Soviet in 1987, providing him with access to the Soviet government's inner circles in a way that NASA administrators lack in the United States. It may well be that the good health of the Soviet space program, and particularly of its scientific component, strongly depends on Sagdeev and his influential position. As this book went to press, Sagdeev quit his position and denounced the Soviet space shuttle as scientifically worthless.

Indeed, there are some disturbing similarities between the Chinese exploration of the Indian Ocean in the century before Columbus and a potential future of the Soviet space program. The Chinese treasure ships traveled far and wide because they brought prestige to the emperor and because the influential bureaucrat Cheng Ho was in a position to press for their continuance. This exploratory program came to a dead stop when Cheng Ho was no longer in the picture. A similar future for the Soviet space program might well occur if their trip to Mars was taken for purely symbolic reasons. Once the hammer and sickle flutters over the sands of Mars, if that's the sole goal of the Soviet space program, why press on?

A MORE OPTIMISTIC HISTORICAL ANALOGY

History can be used to create depressing analogies, but some more optimistic ones are available too. The best type of Martian expedition would be rather similar to our return to the Antarctic in 1957 with the International Geophysical Year (IGY). While scientists originated the concept of the IGY, governments embraced it because it offered an opportunity to do what they wanted to do anyway—launch an earth-orbiting satellite and maintain a national presence in the Antarctic. Launching a satellite for scientific reasons had the virtue of establishing the right to orbit satellites over enemy territory, a right which was later on essential in establishing the reconnaissance satellite program. The economic potential of the Antarctic, while untapped for centuries, was certainly reason for nations interested in the Antarctic to stay there.

Just as there are explorers and scientists who are willing to spend their careers focusing on Mars, so were there many earth scientists who wanted to go to the Antarctic. The problem that both groups face is persuading some government to support a sustained exploration program. The IGY infrastructure continued to exist partly because IGY activities established facilities which could be used for the forseeable future, such as the Amundsen–Scott South Pole Station established by the United States. In addition, the major countries involved in the IGY believed, and continue to believe, that at some time in the future the Antarctic would have considerable geopolitical or economic importance. Thus while science and the exploring spirit motivated the IGY, the Antarctic's potential as a natural resource was an absolute necessity in persuading governments to spend money on the expedition.

With the right institutional context, then, the trip to Mars could indeed be quite parallel to the IGY. The point is not that the trip itself can be justified in an economic way, but rather that it is a good focal point for a group of activities which can, eventually, be self-supporting. Whether or not establishing a colony on the Martian surface makes long-term economic sense, a trip to Mars will probably make sense in the broader context of a 21st century space program. Space enthusiasts must hope that America, the Soviets, or whoever decides to go to Mars has the patience to undertake the expedition as part of a broader push into space rather than as a mad dash which simply leads to the sands of Mars and then nowhere.

List of Acronyms (LOA)

ABM—Antiballistic missile
ASAT—Antisatellite weapons system
BMD—Ballistic missile defense
CBA—Controlled Bureaucratic Ambiguity
CELSS—Controlled Environmental Life Support System, or Controlled Ecological Life Support System, or Closed Ecological Life Support System
CRAF—Comet Rendezvous/Asteroid Flyby
DACT—Disposable absorbent containment trunk
DOD—Department of Defense
EOS—Electrophoresis Operations in Space
ESA—European Space Agency
EURECA—European Retrievable Carrier
EUVE—Extreme Ultraviolet Explorer
EVA—Extravehicular activity
GAS—Get Away Special program
GEO—Geosynchronous (or geostationary) orbit
HST—Hubble Space Telescope
IGY—International Geophysical Year
INTELSAT—International Telecommunications Satellite Corporation
IRAS—Infrared Astronomy Satellite
ISF—Industrial Space Facility
IUE—International Ultraviolet Explorer
JEA—Joint Endeavor Agreement
JPL—Jet Propulsion Laboratory
KH—Keyhole (reconnaissance satellite)
LEO—Low earth orbit
LOA—List of acronyms
MAD—Mutual assured destruction

MBE—Molecular beam epitaxy
MESA—Modular Experimental Platform for Science and Applications
MLR—Monodisperse Latex Reactor
MMU—Manned maneuvering unit
MOL—Manned Orbiting Laboratory
NASA—National Aeronautics and Space Administration
NEAR—Near Earth Asteroid Rendezvous
NSF—National Science Foundation
OAO—Orbiting Astronomical Observatory
Ph-D—Phobos/Deimos mission
SAS—Space adaptation syndrome
SMS—Space motion sickness (also known as SAS)
SDI—Strategic Defense Initiative
SPAS—Space Pallet Satellite
SPOT—Système Probatoire pour l'Observation de la Terre (French satellite that takes high resolution pictures of earth)
SURF—Space Ultravacuum Facility
TDRS—Tracking and Data Relay Satellites
WCS—Waste collection system

Reference Notes

CHAPTER 1

1. Balboa's expedition is described in Samuel E. Morison, *The European Discovery of America: The Southern Voyages* (New York: Oxford University Press, 1974), pp. 200–204.
2. Harry L. Shipman, *Space 2000: Meeting the Challenge of a New Era* (New York: Plenum, 1987).
3. For instance, see Sally K. Ride, *Leadership and America's Future in Space* (Washington, D.C.: NASA, 1987); James Oberg and Alcestis Oberg, *Pioneering Space: Living on the Next Frontier* (New York: McGraw-Hill, 1986); and National Commission on Space, *Pioneering the Space Frontier* (New York: Bantam, 1986).
4. Others, of course, have posed these questions. I found these rather clearly stated in Bruce Murray, Michael C. Malin, and Roland Greeley, *Earthlike Planets: Surfaces of Mercury, Venus, Earth, Moon, Mars* (San Francisco: Freeman, 1981), 350.
5. Spark M. Matsunaga, *The Mars Project* (New York: Hill and Wang, 1986).
6. Shipman, *Space 2000*, pp. 329–332.
7. The $5000 figure is based on the costs of launching payloads with expendable rockets in the 1970s, converted into 1988 dollars; for references, see H. L. Shipman, *Space 2000: Meeting the Challenge of a New Era* (New York: Plenum, 1986), p. 396. The $10,000 figure is based on the full costs of the shuttle program, for which I take the following simple approach: From 1970 through 1995, the space shuttle cost approximately $15 billion to build and $2 billion per year (from 1980 through 1995) to operate, for a total cost of $45 billion in 1988 dollars, in rough numbers, and projecting that operating costs will continue at the current rate. Optimistically, there will be 75 shuttle launches through 1995, with 25 launches from 1980 through 1988 and an average of eight launches per year from 1989 through 1995. With 60,000 pounds being (or capable of being) launched with each shuttle flight, the arithmetic shows that the space shuttle can or will launch a total of $75 \times 60{,}000 = 4.5$ million pounds of stuff into low orbit, corresponding to a launch cost of $10,000 per pound.
8. Paul Kennedy, *The Rise and Fall of the Great Powers* (New York: Random House, 1987).
9. See, for example, William K. Hartmann, Ron Miller, and Pamela Lee, *Out of the Cradle: Exploring the Frontiers Beyond Earth* (New York: Workman Publishing, 1984); G. K. O'Neill, *The High Frontier: Human Colonies in Space* (New York: William Morrow, 1977),

2081: A Hopeful View of the Human Future (New York: Simon and Schuster, 1981); Ben Bova, *The High Road* (Boston: Houghton Mifflin, 1981); and much, but not all, of National Commission on Space, *Pioneering the Space Frontier* (New York: Bantam, 1986).

10. The first 50 years are based on *Pioneering the Space Frontier*, p. 190, and on the Ride report. I think the timetable in both reports is quite optimistic and have adjusted the dates accordingly.
11. James A. Van Allen, "Myths and Realities of Space Flight," *Science 232* (30 May 1986): 1075–1076; letters responding to this are in *Science 233* (8 August 1986), 610–611.
12. Thomas Donahue *et al.*, Study Steering Group, Space Science Board, National Research Council, *Space Science in the Twenty-First Century: Imperatives for the Decades 1995 to 2015: Overview* (Washington, D.C.: National Academy Press, 1988), pp. 78–80.
13. Freeman Dyson, *Disturbing the Universe* (New York: Harper and Row, 1979), Chapt. 11.

CHAPTER 2

1. Quoted in William K. Hartmann, Ron Miller, and Pamela Lee, *Out of the Cradle: Exploring the Frontiers Beyond Earth* (New York: Workman Publishing, 1984), p. 7.
2. Daniel J. Boorstin, *The Exploring Spirit: America and the World, Then and Now* (New York: Random House, 1976).
3. F. Nansen, introduction to Roald Amundsen, *The South Pole*, trans. A. G. Chater (New York: Lee Keedrick, 1913), p. xxix. Emphasis in the original.
4. David Falkner, "Adventurers Busily Explore Final Frontier: Imagination," *The New York Times* (February 10, 1988), pp. A1, D27.
5. Michael Collins, *Carrying the Fire* (New York: Farrar, Straus, and Giroux, 1974), pp. 385–386.
6. One source for videos is The Planetary Society, 65 N. Catalina Avenue, Pasadena, CA 91106. Teachers can obtain free loan films, and can copy videos, from appropriate offices at the NASA field centers like the Johnson Space Center in Houston. The Astronomical Society of the Pacific, 390 Ashton Ave., San Francisco, CA 94122, is another source of visual material.
7. E.E. Aldrin, Jr. with Wayne Warga, *Return to Earth* (New York: Random House, 1973).
8. Joseph P. Allen with Russell Martin, *Entering Space: An Astronaut's Odyssey* (New York: Stewart, Tabori, and Chang, 1985); Bill Nelson with Jamie Buckingham, *Mission: An American Congressman's Voyage to Space* (New York: Harcourt, Brace, Jovanovich, 1988); Sally Ride with Susan Okie, *To Space And Back* (New York: Lothrop, Lee, and Shepard, 1986); William R. Pogue, *How Do You Go to the Bathroom in Space?* (New York: Tor Books, 1985).
9. Tom Wolfe, *The Right Stuff* (New York: Bantam, 1983).
10. Stephen J. Pyne, *The Ice: A Journey to Antarctica* (Iowa City: University of Iowa Press, 1986), especially pp. 170ff.
11. Robert F. Scott, *Scott's Last Expedition: The Personal Journals of Captain R.F. Scott, R.N., C.V.O., on his Journey to the South Pole*, intr. by J.M. Barrie (New York: Dodd, Mead & Co., 1913), p. 162. See also Peter Brent, *Captain Scott* (New York: Saturday Review Press, 1974); Ian Cameron, *Antarctica: the Last Continent* (Boston: Little, Brown & Co., 1967).
12. Scott, *Diary*, pp. 464–465, 470–477.

13. Henry S. F. Cooper, *Thirteen: the Flight that Failed* (New York: Dial Press, 1973), pp. 21–37.
14. Cooper, p. 148.

CHAPTER 3

1. James A. Van Allen, "Myths and Realities of Space Flight," *Science 232* (30 May 1986): 1075–1076; letters responding to this article are given in *Science 233* (8 August 1986): 610–611. I described Van Allen's views briefly in *Space 2000*, pp. 324–327.
2. See Shipman, *Space 2000*, pp. 72ff; and A. C. Clarke, *Ascent into Orbit: A Scientific Autobiography* (New York: Wiley, 1984).
3. The White House, Office of the Press Secretary, Fact Sheet on the President's Space Policy, released February 1988.
4. See, for example, Henri Pirenne, *Economic and Social History of Medieval Europe*, trans. I. E. Clegg (New York: Harcourt, Brace, and World, 1937), pp. 142–146.
5. The role of spices in medieval and Renaissance Europe is described by Frederic Rosengarten, Jr., *The Book of Spices* (New York: Pyramid, 1973), pp. 60–65.
6. *Wage Comparisons:* Jaime Vicens Vives, *An Economic History of Spain*, Frances M. Lopez-Morillas (trans.), (Princeton: Princeton University Press, 1969) writes that 1 *ducat* or 375 *maravedis* paid eight days' wages for a "specialized laborer" (p. 296) at the time of Columbus. Setting this at about $6/hour for an eight-hour day means, in contemporary purchasing power terms, 375 *maravedis* = $384. Don Quixote's aide Sancho Panza earned 26 maravedis per day (Miguel de Cervantes Saavedra, *The Adventures of Don Quixote de la Mancha*, T. Smollett (trans.), (New York: Farrar, Straus, and Giroux, 1986). If you equate this to a minimum wage worker at MacDonald's earning $28 for an eight-hour day, you once again arrive at the figure of 1 *maravedi* = $1.

    Another, more detailed salary comparison can be made using the data from Samuel E. Morison's *The European Discovery of America: The Southern Voyages* (New York: Oxford University Press, 1974). He lists various salaries for different members of the expedition (all figures in *maravedis* per year):

Magellan and Faleiro, joint captains (tripled Magellan's before sailing)	50,000
Luis de Mendoza, ship captain and treasurer of the fleet	60,000
Juan de Cartagena, ship captain and inspector general of fleet	60,000
Pilots (plus a few thousand extra for piloting Magellan)	10,000
Able seaman	14,400
Gromets or apprentice seamen	9,600

    In salary terms, all of the above numbers say that a *maravedi* is something like $1; curiously, by modern standards the pilots and captains are underpaid relative to the seamen. Salaries of members of the Columbus expedition (S. E. Morison, *Admiral of the Ocean Sea*, Boston: Little, Brown & Co., 1942, Vol. 1, p. 137) are similar.

    Cost of Living: Morison's *Admiral of the Ocean Sea* cites a cost of 12 *maravedis* for feeding a seaman for a day; I'd interpret this as meaning in contemporary terms that a *maravedi* is, in terms of food, worth something like 25–50 cents. The cost of maintaining a manservant in Pavia, Italy, was 20 *florins* [Carlo M. Cipolla, *Money, Prices, and Civilization in the Mediterranean World from the 5th Through 17th Centuries* (Princeton: Princeton University Press for the University of Cincinnati, 1956)]. The *florin* (minted in Florence) and the *ducat* (minted in Venice; referred to in Shakespeare's *Merchant of Venice*) were two stable currencies of the time, and worth 375 *maravedis* (Vicens Vives, p. 296). The arithmetic shows that the cost of maintaining such a person is 7500 *maravedis* per year. The person in question was a high-level manservant; I interpret the data as indicating that the purchasing power of a *maravedi* was, again, something like 50 cents.

With the purchasing power equivalent of a *maravedi* being roughly 50 cents, and the salary equivalent being about $1, I'll split the difference and use an equivalent of 1 *maravedi* = 75 cents; as explained in the text, for current purposes we need to interpret such things as the cost of Columbus's expedition and the cost of spices both in terms of goods which were purchased then and now (e.g., food and shelter) and in terms of how many man-years of work is invested in something like 10 pounds of cloves or an oceanic voyage.

Gold equivalents: The usual approach taken by historians is to equate old currencies to gold, and then use contemporary prices to set dollar equivalents. The *florin* and *ducat* were gold coins with weights of 3.5 to 4.5 grams or 0.12 to 0.16 troy ounces (Cipolla, pp. 22–23) and worth 375 *maravedis*, according to the exchange rate quoted by Vicens Vives for the time of Ferdinand and Isabella. Prices often quoted in older history books are based on gold at $35 an ounce, a price fixed for decades by the U.S. government, leading to 1 *maravedi* being worth $0.013, unrealistically low in purchasing power. Gold prices in 1988 ($500 per ounce, roughly speaking) are closer to what they were in the 15th century; using these prices, 1 *maravedi* = 18 cents, still lower than its worth in purchasing power or salary equivalents by a considerable factor.

7. Rosengarten, *The Book of Spices*, gives (p. 205) clove prices as ranging from 33 cents per pound (1967) to $2.25 per pound (1972); cumin (p. 220) as 26 cents per pound; and nutmeg (p. 300) as 40 cents per pound. The approximate figure in the text allows for a threefold price increase since the early 1970s.
8. The price of draft animals is estimated based on my past experience as business consultant to a horse trader and on the 1987 price of feeder cattle (60 cents per pound; a lightweight feeder cow, similar to a poorly fed 15th-century cow, would cost about $600 on the commodity market.
9. See John H. Parry, *The Age of Reconnaissance* (New York: Mentor, 1964), pp. 56–58; Rosengarten, *The Book of Spices*, p. 68; and Daniel Boorstin, *The Discoverers* (New York: Random House, 1983), Chapt. 19.
10. John H. Parry, *The Discovery of the Sea* (New York: Dial, 1974), pp. 85, 100–102.
11. Bibliotheque Nationale, Paris, MS. Espagnol 30, quoted in John H. Parry, *The Age of Reconnaissance*, p. 33.
12. William H. Goetzmann, "Paradigm Lost," in Nathan Reingold (ed.), *The Sciences in the American Context: New Perspectives* (Washington, D.C.: Smithsonian Press, 1979), pp. 21–34; Goetzmann, *New Lands, New Men: America and the Second Great Age of Discovery* (New York: Viking, 1986).
13. The full story of Shackleton's expedition is told in E. Shackleton, *South*, (New York: MacMillan, 1920). The open boat voyage is told in an extract in Sir Francis Chichester, *Along the Clipper Way* (New York: Ballantine, 1966), pp. 239–260.
14. Paul C. Daniels, *The Antarctic Treaty*, in Richard S. Lewis and Philip M. Smith (eds.), *Frozen Future: A Prophetic Report from Antarctica* (New York: Times Books, 1973), p. 34.
15. Pyne, p. 350.

16. Paul C. Daniels, "The Antarctic Treaty," in *Frozen Future*, p. 35.
17. *Frozen Future*, p. 58.
18. Pyne, *The Ice*, p. 352.
19. Boorstin, *Discoverers*, p. 192.
20. Boorstin, *Discoverers*, p. 201.

CHAPTER 4

1. Lyndon Baines Johnson, *The Vantage Point: Perspectives of the Presidency 1963–1969* (New York: Holt, Rinehart, and Winston, 1971), p. 272.
2. See, for example, Dwight David Eisenhower, *Waging Peace 1956–1961: The White House Years* (New York: Doubleday, 1965), p. 206; W. Von Braun, F. I. Ordway III, and Dave Dooling, *Space Travel: A History* (New York: Harper and Row, 1985), p. 170.
3. Sally K. Ride, *Leadership and America's Future in Space* (Washington, D.C.: NASA, 1987), pp. 16–19.
4. Tables 2 and 3 are taken largely from the Ride report, p. 19; I've added the Skylab mission and left out quite a few.
5. H. L. Shipman, *Space 2000: Meeting the Challenge of a New Era* (New York: Plenum, 1987), pp. 97–99.
6. This story has been told many times, most notably in F. I. Ordway III and M. R. Sharpe, *The Rocket Team* (Cambridge, Mass.: M.I.T. Press, 1982), and in W. Von Braun, F. I. Ordway III, and Dave Dooling (collaborating author), *Space Travel: A History* (New York: Harper and Row, 1985), pp. 121–123.
7. Walter A. MacDougall, . . . *the Heavens and the Earth* (New York: Basic Books, 1985), p. 122.
8. See, for example, Shipman, *Space 2000*, p. 24.
9. See MacDougall, pp. 120ff; J. A. Van Allen, *The Origins of Magnetospheric Physics* (Washington, D.C.: Smithsonian Institution, 1983), pp. 33–59.
10. For the details of the launch decision, see *Space 2000*, Chapter 1, as well as the *Report to the President by the Presidential Commission on the Space Shuttle Challenger Accident* (Washington, D.C.: Government Printing Office, 1986).
11. John Logsdon, *The Decision to Go to the Moon* (Cambridge: MIT Press, 1970).
12. A brief history of the supercomputer initiative is in S. Karin and N. P. Smith, *The Supercomputer Era* (Boston: Harcourt, Brace, Jovanovich, 1987), p. 106.
13. The role of NASA is discussed by MacDougall, p. 317.

CHAPTER 5

1. Sally K. Ride with Susan Okie, *To Space and Back* (New York: Lothrop, Lee, and Shepard, 1986), p. 29.
2. Joseph P. Allen with Russell Martin, *Entering Space: An Astronaut's Odyssey* (New York: Stewart, Tabori, and Chang, rev. ed., 1985), pp. 81–83.
3. William R. Pogue, *How Do You Go to the Bathroom in Space?* (New York: Tor Books, 1985), p. 29.

4. Joseph Allen, *Entering Space*, p. 75.
5. Pogue, *How Do You Go to the Bathroom in Space?*, p. 29.
6. Life Sciences Division, NASA Office of Space Science and Applications, *Life Sciences Accomplishments* (Washington, D.C.: NASA, 1986), pp. 4–6.
7. A. E. Nicogossian and J. F. Parker, *Space Physiology and Medicine*, NASA SP-447 (Washington, D.C.: NASA, 1982), Chapt. 8; Ride, *To Space and Back*, p. 32.
8. Ride, *To Space and Back*, p. 42.
9. James and Alcestis Oberg, *Pioneering Space: Living on the Next Frontier* (New York: McGraw-Hill, 1986), p. 178.

CHAPTER 6

1. *The CIBA Collection*, Vol. 8: *The Musculoskeletal System*, Vol. 1, pp. 178–183. I thank Mary Whitmore of the University of Oklahoma for calling my attention to this article.
2. James E. Oberg and Alcestis R. Oberg, *Pioneering Space* (New York: McGraw-Hill, 1986), pp. 130–131.
3. Arnauld E. Nicogossian and James F. Parker, *Space Physiology and Medicine*, NASA SP-447 (Washington, D.C.: NASA, 1982), p. 207.
4. Nicogossian and Parker, p. 206.
5. Felicity Barringer, "Soviet Astronaut Says Exercise Aided Adjustment to Gravity," *New York Times* (January 21, 1988), p. A22.
6. Mary M. Connors, Albert A. Harrison, and Faren R. Akins, *Living Aloft: Human Requirements for Extended Spaceflight*, NASA SP-483 (Washington, D.C.: NASA, 1985), p. 83.
7. Connors, Harrison, and Akins, p. 82.
8. The magnitude of the acceleration produced by the centrifugal force is just v^2/r, where v is the velocity (in, say, meters per second) of the rim of the station and r is its distance from the center. If the station or craft rotates once every T seconds, you calculate the force by using the fact that $2 \pi r = vT$ or $v=2 \pi r/T$, making the acceleration equal to $(2\pi r/t)^2/r$ or, cancelling terms, $(2\pi)^2 r/T^2$. Setting this acceleration equal to the acceleration of gravity on earth, 9.8 m/s^2, and manipulating a bit produces a remarkably simple relation between spacecraft radius r and revolution period T: T^2 (in seconds) $= 4 R$ (in meters).
9. Richard D. Johnson and Charles Holbrow, in *Space Settlements: A Design Study*, NASA SP-413 (Washington, D.C.: NASA, 1977), list a number of proposed habitat systems on p. 42; systems about 30 m in radius have been mentioned previously. I don't know of anyone who has addressed the question of a minimally sized space station with artificial gravity in any real detail.
10. Doing the arithmetic, 15 pounds per square inch corresponds to about 10 tons (metric tons) per square meter. Modeling our putative space colony as a 6-m diameter cylinder which is 36 m long, I get a surface area (including the ends) of 736 square meters, thus requiring 7360 tons of rock or, rounding off, about 250 space shuttle flights with 30 tons per flight.
11. The surface area of this storm cellar is 80 m^2, almost a factor of 10 smaller than the space station modeled earlier. Substituting 0.2 m of aluminum for 2 m of rock saves another factor of 10, accounting for the tremendous difference.
12. Nicogossian and Parker, p. 298.
13. Frank Miles and Nicholas Booth (general editors), *Race to Mars: The Mars Flight Atlas* (New York: Harper and Row, 1988), pp. 72–73.

14. Mary M. Connors, Albert A. Harrison, and Faren R. Akins, *Living Aloft: Human Requirements for Extended Spaceflight*, NASA SP-483 (Washington, D.C.: NASA, 1985).
15. Robert L. Helmreich, John A. Wilhelm, and Thomas E. Runge, "Psychological Considerations in Future Space Missions," in T. Stephen Cheston and David L. Winter (eds.), *Human Factors of Outer Space Production* (Boulder, Colo.: Westview Press, 1980), AAAS Selected Symposium 590, pp. 1–24; Roland Radloff and Robert Helmreich, *Groups Under Stress: Psychological Research in Sealab II* (New York: Appleton-Century Crofts, 1968).
16. Ian Cameron, *Antarctica: The Last Continent* (Boston: Little, Brown, 1974), pp. 146–159.
17. James E. Oberg and Alcestis R. Oberg, *Pioneering Space: Living on the Next Frontier* (New York: McGraw-Hill, 1986), pp. 194–195.
18. Kirmach Natani, "Future Directions for Selecting Personnel," in T. Stephen Cheston and David L. Winter (eds.), *Human Factors of Outer Space Production* (Boulder, Colo: Westview Press, 1980), pp. 25–63.
19. Shipman, *Space 2000*, pp. 332–335.
20. Helmreich, Wilhelm, and Runge, p. 16.

CHAPTER 7

1. The numbers are from H.L. Shipman, *Space 2000: Meeting the Challenge of a New Era* (New York: Plenum, 1987), p. 329, and from the *Life Sciences Report: December 1987*, prepared by NASA's Office of Space Science and Applications, Life Sciences Division (Washington, D.C.: NASA, 1987), p. 44.
2. Some conceptions of space food are provided by Joe Allen and Russell Martin, *Entering Space: An Astronaut's Odyssey* (New York: Steward, Tabori, and Chang, 1984), pp. 75–77; James and Alcestis Oberg, *Pioneering Space* (New York: McGraw-Hill, 1986), pp. 184–187.
3. Oberg and Oberg, *Pioneering Space*, p. 184.
4. Harlan F. Brose, "Environmental Control and Life Support (ECLS) Design Optimization Approach," in Mireille Gerard and Pamela Edwards (eds.), *Space Station: Policy, Planning, and Utilization* (New York: American Institute of Aeronautics and Astronautics, 1983), pp. 189–194.
5. Brose, p. 191; Oberg and Oberg, Chapt. 6.
6. R. L. Sauer, "Metabolic Support for a Lunar Base," in W. W. Mendell (ed.), *Lunar Bases and Space Activities of the 21st Century* (Houston: Lunar and Planetary Institute, 1985), pp. 647–652.
7. Space Station Reference Configuration Group, *Engineering and Configurations of Space Stations and Platforms* (Park Ridge, N.J.: Noyes Publications, 1985), p. 470.
8. *Engineering and Configurations of Space Stations and Platforms,* p. 464.
9. Life Sciences Division, NASA Office of Space Science, *Life Sciences Accomplishments: December 1986*, and *Life Sciences Report: December 1987*, both published at Washington, D.C. by NASA. For copies contact Code EB, Life Sciences Division, NASA Headquarters, Washington, D.C. 20546.
10. Oberg and Oberg, *Pioneering Space*, Chapt. 8.
11. *New York Times* (24 June 1986): C3.
12. T.D. Lin, "Concrete for Lunar Base Construction," in W. W. Mendell (ed.), *Lunar Bases and Space Activities of the 21st Century* (Houston: Lunar and Planetary Institute, 1985), pp. 381–390.

13. Peter Smolders, *Living in Space* (New York: Ballantine, 1982), p. 2.2; Brose, p. 192; *Engineering and Configurations of Space Stations and Platforms*, p. 466.

CHAPTER 8

1. Numbers are from Harry L. Shipman, *Space 2000: Meeting the Challenge of a New Era* (New York: Plenum, 1987), Chapt. 3; John S. Lewis and Ruth A. Lewis, *Space Resources: Breaking the Bonds of Earth* (New York: Columbia University Press, 1987), p. 169; Paul Keaton, "A Moon Base/Mars Base Transportation Depot," in W. W. Mendell (ed.), *Lunar Bases and Space Activities of the 21st Century* (Houston: Lunar and Planetary Institute, 1985), pp. 141–154.
2. To follow the straight-line path, radially outward from the sun, from earth to Mars requires you to put on the brakes, getting rid of the earth's velocity around the sun (velocity change of 30 km/sec), and to match velocities with Mars, requiring you to accelerate to a speed of 22 km/sec around the sun, giving a 52 km/sec from an earth-escape trajectory to orbit around Mars. You then need to add what it takes to get from LEO to earth-escape, to get into orbit around Mars, and some suitable acceleration to push you on this impossible path.
3. If the mass of the fuel is F and the mass of the payload is P, then the rocket equations given by Keaton, "Moon/Mars Transportation Depot" (p. 145), give $F/(P+F) = [1 - e^{-\delta V/c})]$ where $e = 2.718282$, the base of natural logarithms, and c is the rocket exhaust velocity. If F is much bigger than P, this expression can be simplfied to give an approximate relation $F/P \sim e^{\delta V/c})$.
4. The mission profile is taken from Kerry Mark Joels, *The Mars One Crew Manual* (New York: Ballantine Books, 1985). Joels doesn't give exact weights, and makes some assumptions about space shuttle improvements. In his mission profile, established before the *Challenger* aftermath made it clear that using the space shuttle to cart fuel into orbit is both dumb and dangerous, 28 space shuttle flights plus five flights of a hypothetical "G1" Soviet booster plus six Ariane flights were required to put the "Mars One" equipment in orbit. Assume that this hypothetical "G1" booster, with a hypothetical capacity of 400,000 pounds to low earth orbit, is roughly like the old Saturn 5 [which weighed 6.4 million pounds at launch and could put 300,000 pounds into low orbit, according to W. Von Braun and F.I. Ordway III, *History of Rocketry and Space Travel* (New York: Crowell, 1969), pp. 150-170)]; these capabilities are roughly similar to the USSR's recently flown ENERGIA. This sets the launch weight of the G1 at 8.5 million pounds. Add up 28 shuttles at 4.4 million pounds each and six G1's at 8.5 million pounds each and you get 163 million pounds of fuel required for the mission; the remaining weight is for the spacecraft, people, and consumables.

   Of course there are many ways to accomplish a Mars mission, and since much of the machinery isn't designed yet, numbers like these should only be considered illustrative. I bury this calculation in a footnote to show that the very rough numbers which I cite in the text are not that far off what may become reality, *if* we use chemical rockets and nothing but chemical rockets to send people to Mars.
5. Neither ion engines nor solar sails are mentioned in the NASA Space Systems Technology Model, NASA TM 88174, a compilation of the status of various NASA programs. Nuclear-powered ion engines are mentioned briefly as a program with much past work but little for the immediate future in the Committee on Advanced Space Technology's report, *Space*

Technology to Meet Future Needs (Washington, D.C.: National Academy of Sciences, 1987), Chapt. 5.
6. NASA, Office of Aeronautics and Space Technology, *Project Pathfinder: Research and Technology to Enable Future Space Missions* (Washington, D.C.: NASA, December 1987).
7. Stan Kent, "Solar Electric Propulsion Stage as a Mars Exploration Tool," in P. Boston (ed.), *The Case for Mars* (San Diego: Univelt, 1981), 83–91.
8. For information on ion engines, see Louis Friedman, *Starsailing: Solar Sails and Interstellar Travel* (New York: Wiley, 1988), Chapt. 5; Lewis and Lewis, *Space Resources*, pp. 143–147.
9. *Space Technology to Meet Future Needs,* pp. 55–56.
10. These numbers are from a detailed study of solar sails, done at the Jet Propulsion Lab in the late 1970s, for a Halley's Comet mission. The study is reported in Louis Friedman, *Starsailing*, Chapts. 2–4.
11. Friedman, *Starsailing*, pp. 135–156.
12. The article is Dana Rotegard, "The Development of Space: The Economic Case for Mars," privately circulated. The references are: (1) Bruce Cordell, "The Moons of Mars: A Source of Water for Lunar Bases and LEO," in W. W. Mendell (ed.), *Lunar Bases and Space Activities of the 21st Century* (Houston: Lunar and Planetary Institute, 1985), pp. 809–818; (2) Robert Farquhar, Goddard Space Flight Center, unpublished paper; (3) Paul Keaton, "A Moon Base/Mars Base Transportation Depot," in *Lunar Bases*, pp. 141–154; (4) Brian O'Leary, "Phobos and Deimos as Resource and Exploration Centers," in C. F. McKay (ed.), *The Case for Mars II* (San Diego: Univelt, for the American Astronautical Society, 1985); (5) A. Sergevetsky, Jet Propulsion Laboratory, unpublished data. I thank Dana Rotegard and Jeff Beddow for their relentless work ferreting out delta-vee data.

CHAPTER 9

1. See, for cxample, L. Don Leet and Sheldon Judson, *Physical Geology*, 4th ed. (Englewood Cliffs, NJ: Prentice-Hall, 1971), pp. 123–130.
2. J. H. Alton, C. Galindo, Jr., and L.A. Watts, "Guide to Using Lunar Soil and Simulants for Experimentation," in W. W. Mendell (ed.), *Lunar Bases and Space Activities of the 21st Century* (Houston: Lunar and Planetary Institute, 1985), pp. 497–506.
3. R. Silberberg, C. H. Tsao, J. H. Adams, Jr., and J. R. Letaw, "Radiation Transport of Cosmic Ray Nuclei in Lunar Material and Radiation Doses," in Mendell, *Lunar Bases*, pp. 663–669.
4. The cost estimates are from James L. Carter, "Lunar Regolith Fines: A Source of Hydrogen," in Mendell, *Lunar Bases*, p. 71; John S. Lewis and Ruth A. Lewis, *Space Resources: Breaking the Bonds of Earth* (New York: Columbia University Press, 1987), p. 196.
5. This scenario has been discussed by a number of people, including Lewis and Lewis, *Space Resources*, p. 196; David Buden and Joseph Angelo, Jr., "Nuclear Energy—Key to Lunar Development," in Mendell, *Lunar Bases*, pp. 85–98.
6. Lewis and Lewis, *Space Resources*, p. 197.
7. T. D. Lin, "Concrete for Lunar Base Construction," and J. F. Young, "Concrete and Other Cement-Based Composites for Lunar Base Construction," in Mendell, *Lunar Bases*, pp. 381–397.
8. See, for instance, James D. Burke, "Merits of a Lunar Polar Base Location," in Mendell, *Lunar Bases* (Houston: Lunar and Planetary Institute, 1985), pp. 77–85; Carter, "Lunar

Regolith Fines," in Mendell, *Lunar Bases*, pp. 571–581. See also National Commission on Space, *Pioneering the Space Frontier* (New York: Bantam, 1986), p. 140; Solar System Exploration Commitee, *Planetary Exploration through the Year 2000: An Augmented Program* (Washington, D.C.: NASA, 1986), p. 155.
9. Lewis and Lewis, *Space Resources*, pp. 187–188.
10. A brief description of this mission is provided by the Solar System Exploration Committee, *Planetary Exploration through Year 2000: A Core Program* (Washington, D.C.: NASA, 1983), pp. 96–97, 102–104.
11. M. Mitchell Waldrop, "A Soviet Plan for Exploring the Planets," *Science 228*: 698–699 (10 May 1985).
12. Novosti Press Agency, "The Soviet Programme of Space Exploration for the Period Ending in the Year 2000: Plans, Projects, and International Cooperation," background paper distributed in Moscow, October 1987. I thank Norman Ness for providing me with this material.

CHAPTER 10

1. J. Billingham, W. Gilbreath, and B. O'Leary (eds.), *Space Resources and Space Settlements*, NASA SP-428 (Washington, D.C.: NASA, 1979). See particularly the papers in part IV, written by David Bender, R. Scott Dunbar, Michael J. Gaffey, Eleanor Helin, Brian O'Leary, David J. Ross, and Robert Salkeld, in various combinations.
2. Numbers, asteroidal properties, and other information about the asteroids come from various articles in T. Gehrels (ed.) and M. S. Mathews (asst ed.), *Asteroids* (Tucson: University of Arizona Press, 1980).
3. M. J. Gaffey and T. B. McCord, in Gehrels and Mathews, *Asteroids*, p. 701.
4. William K. Hartmann, "The Resource Base in Our Solar System," in Ben R. Finney and Eric M. Jones (eds.), *Interstellar Migration and the Human Experience* (Berkeley: University of California Press, 1985), pp. 26–42.
5. See Clark Chapman's and Laurel Wilkening's chapters in Gehrels and Mathews, *Asteroids* (pp. 25–77).
6. This is a preliminary designation, indicating the year of the designation (1982), the half-month of the discovery (thus "D" designating the last half of February), and a sequence of discovery within that half month. So 1982DB is the second asteroid discovered during the last half of February 1982. A description of asteroidal nomenclature is given by Tom Gehrels, "The Asteroids: History, Surveys, Techniques, and Future Work," in Gehrels and Mathews, *Asteroids*, pp. 3–24.
7. B. O'Leary, "Phobos and Deimos as Resource and Exploration Centers," in C. P. McKay, ed., *The Case For Mars II* (San Diego: Univelt, 1984), pp. 225–244.
8. John S. Lewis and Ruth A. Lewis, *Space Resources: Breaking the Bonds of Earth* (New York: Columbia University Press, 1987), pp. 256–266; A. J. Phillips, "Platinum," *World Book Encyclopedia 15*:502, 1984.
9. Lewis and Lewis, pp. 260–262.
10. Solar System Exploration Committee, *An Augmented Program*, p. 159.
11. Novosti Press Agency, Institute of Space Research, USSR Academy of Sciences, "The Soviet Programme of Space Exploration for the Period Ending in the Year 2000: Plans, Projects, and International Cooperation," background paper distributed in Moscow, 1987. I thank Norman Ness for providing me with this material.

12. European Space Agency, *European Space Science: Horizon 2000*, ESA SP-1070 (Paris: European Space Agency, 1984), p. 10.
13. Solar System Exploration Committee, *An Augmented Program*, pp. 159–173.

CHAPTER 11

1. David W. Smith of the University of Delaware's School of Life and Health Sciences first used this phrase in an extraterrestrial life course that we team-taught.
2. Michael H. Carr, *The Surface of Mars* (New Haven: Yale University Press, 1981), pp. 2–3; Victor R. Baker, *The Channels of Mars* (Austin: University of Texas Press, 1982), p. 4.
3. C. Sagan, *The Cosmic Connection* (Garden City, N.Y.: Doubleday, 1973), p. 130.
4. For a thorough discussion of the life detection experiments, see H. L. Shipman, *Space 2000: Meeting the Challenge of a New Era* (New York: Plenum, 1987), pp. 210–215; Harold P. Klein, "The Search for Life on Mars," in Carr, *The Surface of Mars*, pp. 190–196.
5. Victor R. Baker, *The Channels of Mars* (Austin: University of Texas Press, 1982), pp. 81–85.
6. Baker, *The Channels of Mars*, Chapt. 7; Steven W. Squyres, "The History of Water on Mars," *Annual Review of Earth and Planetary Sciences 12*:83–106, 1984.
7. Michael H. Carr, *The Surface of Mars*, p. 181.
8. S. W. Squyres, "Water on Mars," *Bulletin of the American Astronomical Society 20*:686–687, 1988.
9. S. W. Squyres, "The History of Water on Mars," *Annual Review of Earth and Planetary Science 12*:83–106, 1984.
10. *Ibid.*, p. 103.
11. James B. Pollack, "Atmospheres of the Terrestrial Planets," in J. Kelly Beatty, B. O'Leary, and A. Chaikin (eds.), *The New Solar System*, 2nd ed. (Cambridge, U.K.: Cambridge University Press, and Cambridge, Mass.: Sky Publishing Corporation, 1982), pp. 57–70.
12. K. K. Turekian and S. P. Clark, "Nonhomogeneous Accumulation Model for Terrestrial Planet Formation and Consequences for the Atmosphere of Venus," *Journal of the Atmospheric Sciences* 32:1257-1261, 1975, cited in M. H. Carr, *The Surface of Mars*, p. 186.
13. Joseph Priest, *Energy for a Technological Society*, 2d. ed. (Reading, Mass.: Addison-Wesley, 1979), p. 16.
14. Aleksandr Zakharov, "Close Encounters with Phobos," and Stuart J. Goodman, "Making Tracks on Mars," *Sky and Telescope 76*:17–21, July 1988.

CHAPTER 12

1. Irwin Goodwin, "Beyond INF Treaty, Summiteers Stumble Over Mars Trip and Basic Science," *Physics Today* (July 1988), pp. 47-50.
2. John S. Lewis and Ruth A. Lewis, *Space Resources: Breaking the Bonds of Earth* (New York: Columbia University Press, 1987), p. 245.

CHAPTER 13

1. European Space Agency, *With an Eye to the Future: ESA General Studies Programme 1988*, ESA SP-1100 (April 1988), (Paris: European Space Agency, 1988).
2. "Soviet Commercial Marketing Focuses on Microgravity Flight Opportunities," *Aviation Week and Space Technology* (July 25, 1988), pp. 48–49.
3. See H. L. Shipman, *Space 2000: Meeting the Challenge of a New Era* (New York: Plenum, 1987), Chapt. 4; Peter Brunt and Alan Naylor, "Telecommunications and Space," in Michael Schwarz and Paul Stares, *The Exploitation of Space* (London: Butterworth, 1982), pp. 77–94.
4. William Burrows, *Deep Black: Space Espionage and National Security* (New York: Random House, 1986), pp. 247ff.
5. A good article on Geostar's operations is "Space Operations Begin Using Geostar Payload," by Theresa M. Foley, *Aviation Week* (July 25, 1988), pp. 55–56.
6. Shipman, *Space 2000*, Chapt. 15.
7. Of course, the phrase "catbird seat" is from the memorable sportscaster Red Barber.
8. James J. Haggerty, *Spinoff 85* (Washington, D.C.: Government Printing Office), pp. 36–37; see also R. E. Halpern, in G. A. Hazelrigg and J. M. Reynolds, eds., *Opportunities for Academic Research in a Low-Gravity Environment, Progress in Astronautics and Aeronautics*, Vol. 108 (New York: AIAA, 1986), pp. 28–29; "Space Processed Latex Spheres Sold," *Aviation Week and Space Technology* (22 July 1985), p. 22.
9. Haggerty, *Spinoff 85*, p. 37.
10. Ray A. Williamson, "Tne Industrialization of Space: Prospects and Barriers," in Schwarz and Stares, *The Exploitation of Space*, pp. 70–71.
11. James A. Graham, "Comments on Metals and Alloys," in Schwarz and Stares, pp. 159–163. In *Space 2000* (pp. 145–146) I noticed some confusion in the published literature regarding whether Deere was continuing its experiments; since Graham works for Deere, I regard his statement that Deere is continuing to work with NASA as definitive.
12. T. M. Donahue *et al.* (Study Steering Group), *Space Science in the Twenty First Century: Imperatives for the Decades 1995 to 2015—Overview* (Washington, D.C.: National Academy of Sciences, 1988), pp. 57–59; see also the report of the Fundamental Physics and Chemistry Panel of the same group, chaired by R. Weiss and J. M. Reynolds.
13. "Houston Physicist Gains Ground on Space Vacuum Research Facility," in *Computers in Physics* (Jan./Feb. 1988), pp. 14–15.
14. Shipman, *Space 2000*, pp. 323–324.
15. Gail Bronson, "Mission Irrelevant," *Forbes* (24 March 1986), pp. 176–177.
16. Shipman, *Space 2000*, Chapt. 7.
17. A. R. Oberg, "The Grass Roots of the Mars Conference," in P. J. Boston, ed., *The Case for Mars* (San Diego: Univelt, 1981), p. xi; Solar System Exploration Committee, *Planetary Exploration through the Year 2000: An Augmented Program*, pp. 152–181.

CHAPTER 14

1. Gerald M. Steinberg, "The Militarization of Space," in Michael Schwarz and Paul Stares (eds.), *The Exploitation of Space* (London: Butterworth, 1983), pp. 31–49.
2. William E. Burrows, *Deep Black: Space Espionage and National Security* (New York: Random House, 1986), Chapts. 1–4.

3. Burrows, *Deep Black,* p. 68.
4. See Shipman, *Space 2000: Meeting the Challenge of a New Era* (New York: Plenum, 1987), pp. 102–103 and references cited therein. I was also surprised to find the resolution of high-altitude cameras mentioned in a book for 7-year-olds. Gina Ingoglia and George Guzzi's *The Big Book of Real Airplanes* (New York: Grosset and Dunlap, 1987) cites the resolution of the cameras on the aircraft SR-71A as being able to focus on a golf ball size object on the ground from a flight path 15 miles high. (This book has no page numbers but the citation is seven pages from the end.)
5. Ralph King, Jr., "Eye in the Sky," *Forbes* (December 1, 1986), pp. 216–217.
6. Burrows, *Deep Black*, p. 247.
7. Shipman, *Space 2000*, p. 103.
8. Burrows, *Deep Black,* p. 154.
9. United States Congress, Office of Technology Assesment, *Ballistic Missile Defense Technologies*, OTA-ISC-254 (Washington, D.C.: U.S. Government Printing Office, September 1985), pp. 286–289.
10. William J. Broad, "Beyond the Bomb: Turmoil in the Labs," *New York Times Magazine* (9 October 1988), pp. 23–93 passim.
11. C.K.N. Patel of AT&T Bell Laboratories headed a blue-ribbon panel of scientists to evaluate the Star Wars program. Their report appears in summary form in *Physics Today* (May 1987), pp. S1–S22.
12. Fred Reed, "Hawking Nuclear Snake Oil," *Harper's* (May 1986), pp. 39–48.
13. Yonas is quoted by Robert Scheer in *The Los Angeles Times* (22 September 1985), pp. 1, 14; see also Yonas's article in *Physics Today* (June 1985), pp. 24–32. Abrahamson is quoted by D. Walker, J. Bruce, and D. Cook, "SDI: Progress and Challenges," unclassified staff report submitted to Senators Proxmire, Johnston, and Chiles, p. 12.
14. *The New York Times* (9 October 1988), p. 1.
15. Freeman Dyson discusses just such a scenario in *Weapons and Hope* (New York: Harper and Row, 1984), Chapt. 22.
16. Burrows, *Deep Black,* pp. 276–277.
17. Burrows, *Deep Black*, p. 280.
18. Paul Stares and John Pike, in Schwarz and Stares, *Exploitation of Space*, pp. 110–122; M. Jahani, in Schwarz and Stares, pp. 95–109.
19. Burrows, *Deep Black,* p. 247.
20. Alex Roland, "The Shuttle:Triumph or Turkey," *Discover* (6 November 1985), pp. 29 ff.; Alex Roland, "The Space Shuttle Program: A Policy Failure?" *Science 233* (10 May 1986), pp. 1099-1106.
21. Philip P. Chandler, Leonard David, and Courtland S. Lewis, "MOL, Skylab, and Salyut," in Theodore R. Simpson (ed.), *The Space Station: An Idea Whose Time Has Come* (New York: IEEE Press, 1985), pp. 31–49.
22. Steinberg, "The Militarization of Space," in Schwarz and Stares, *The Exploitation of Space*, p. 42.
23. See footnote 7, chapter 1. With a launch cost of $10,000 per pound, and a carrying capacity of 60,000 pounds, the cost of a shuttle launch, paying for the full cost of developing and maintaining the system, is about $600 million, which I round off to $0.5 billion.
24. Amanda Lee Moore, "Legal Responses for Lunar Bases and Space Activities in the 21st Century"; Christopher C. Joyner and Harrison H. Schmitt, "Extraterrestrial Law and Lunar Bases: General Legal Principles and a Particular Lunar Regime Proposal (Interlune)," both

in Wendell Mendell (ed.), *Lunar Bases and Space Activities of the 21st Century* (Houston: Lunar and Planetary Institute, 1985), pp. 736–749; W. A. McDougall, . . . *the Heavens and the Earth* (New York: Basic Books, 1985), pp. 415–420.
25. Joyner and Schmitt, p. 742.
26. McDougall, . . . *the Heavens and the Earth,* p. 419.
27. I thank Dick Henry of Johns Hopkins University for some entertaining and informative discussions on the military value of Phobos and Deimos.

CHAPTER 15

1. H. L. Shipman, *Space 2000: Meeting the Challenge of a New Era* (New York: Plenum, 1987), Chapt. 12; Don Goldsmith and Tobias Owen, *The Search for Life in the Universe*, (Menlo Park, California: Benjamin/Cummings, 1980).

CHAPTER 16

1. The White House, Office of the Press Secretary, press release dated January 26, 1988.
2. European Space Agency, *European Space Science: Horizon 2000*, ESA SP-1070 (Paris: European Space Agency, 1984).
3. Stephen B. Hall (ed.), *The Human Role in Space: Technology, Economics, and Optimization* (Park Ridge, NJ: Noyes Publications, 1985).
4. The calculation is done most easily if we assume that this facility will exist for a substantial fraction of the useful life of the United States space station, which for accounting purposes can be taken to be ten years—a long estimate, leading to low costs; the usual lifetimes for high-technology equipment are five years, not ten. Even using NASA figures for the cost of the eight-person space station, which are probably low, SVT will need to invest $2 billion (1/8th of $16 billion) just to house the astronaut.
5. At current, fully priced shuttle launch costs (defined in Chapter 14) the cost of launching an astronaut, priced at 1/7 of a shuttle mission, is about $80 million. Even if a relief person is only sent up once a year, the launch cost is $80 million per year to launch this one-person crew into space, or $0.8 billion over the ten-year period. The share of the cost of keeping this person alive comes from a requirement of 20 pounds per day of consumables (including cooling water for the spacecraft); I'm assuming that consumables are launched by expendable rockets, considerably cheaper than the shuttle at a cost of $3000 per pound. These numbers are only ballpark figures.
6. For descriptions of these platforms, see H. L. Shipman, *Space 2000:Meeting the Challenge of a New Era* (New York: Plenum, 1987), Chapt. 14.
7. See W. Goetzmann, *New Lands, New Men: America and the Second Great Age of Discovery* (New York: Viking 1986); see also Carl Sagan and Stephen J. Pyne, *The Scientific and Historical Rationales for Solar System Exploration*, publication SPI 88-1 (Washington, D.C.: Space Policy Institute, The George Washington University, Washington, DC 20052, 1988).
8. European Space Agency, *Horizon 2000*, p. 3.
9. Concrete evidence of the salability of these stories is shown in their appearance in "Physics News in 1983, 1984, 1985, . . ."; a section of the January issue of *Physics Today* in 1984,

1985, and 1986, respectively. Stories falling outside these categories which I suggested for "Physics News," and for which I prepared press releases on other occasions, generally were not used.

10. Currently, space science is supported at a level of about $2 billion per year, which is about two thirds of what the entire United States spends on professional sports ($3.1 billion in 1988, according to the *Standard and Poor Industry Surveys* (p. L16) or on theater, opera, and symphonies combined. In other words, for space science to hold its own, we'd have to garner a share of the entertainment dollar which is comparable to those two activities. This amount of money would have to go exclusively to space science, and space scientists would be competing with scientists in other fields for the same pool of resources.

    Another indication of the difficulty of selling a pure science space program is the reaction of various groups of science supporters to the vocal arguments of some space scientists that we should abandon the manned space program entirely. *Astronomy* is a magazine for amateur astronomers, and if any significant segment of the general public could be expected to support a purely unmanned space science program, it is *Astronomy*'s readers. Only a minority echoed the view that unmanned space flight is the space program's most important goal. See Kristine R. Majdacic, "Shaping America's Future in Space," *Astronomy* (May 1988), pp. 16–17.
11. Clearly all those on the Forbes 500 list of the wealthiest Americans, all of whom have net worths well in excess of $100 million, could afford million-dollar vacations in space if they wanted them.
12. If you estimate that the probability of obtaining water in space is W, the probability of successful industrialization is C, and the probability that tourism makes economic sense is T, then the likelihood of each scenario works out to be, where each probability is a decimal: space settlement = WC; space industrialization = $(1-W)C$; research only = $(1-W)(1-C) + W(1-C)(1-T)$; tourism = $WT(1\text{-}C)$. The numbers in the text use my probabilities of W=0.7, C=0.6, and T=0.5.

CHAPTER 17

1. For a general discussion of Mars missions, see John Butler, "Mission and Space Vehicle Concepts," in M.B. Duke *et al., Manned Mars Missions: Working Group Papers* (Washington, D.C.: NASA, 1986), NTIS Documents N87-17722 through N87-17759.
2. Kerry Mark Joels, *The Mars One Crew Manual* (New York: Ballantine Books, 1985); see also Gus R. Babb and William R. Stump, "Comparison of Mission Design Options for Manned Mars Missions," in Duke *et al., Manned Mars Missions: Working Group Papers*, pp. 162–188.
3. National Commission on Space, *Pioneering the Space Frontier* (New York: Bantam Books, 1986), p. 136.
4. Space Science Board, National Academy of Sciences, *Space Science in the 21st Century: Imperatives for the Decades 1995–2015*, (Washington, D.C.: National Academy Press, 1988); see also Craig Covault, "Science Board Proposes New Space Program Direction," *Aviation Week* (August 1, 1988), pp. 36–40.
5. I'm writing this in the summer of 1988, at one of those times which recur at two-year intervals when Mars rises around sunset, is high in the south at midnight, and sets in the west around sunrise.

6. Spark Matsunaga, *The Mars Project* (New York: Hill and Wang, 1986).
7. I'm using a world population of about 5.3 billion people and a combined population for the United States, the USSR, Japan, and Western Europe of 1 billion.
8. National Commission on Space, *Pioneering the Space Frontier*, pp. 175–180.
9. For example, Leon Jaroff (with Glenn Garelik, J. Madeliene Nash, and Richard Woodbury), "Onward to Mars," *Time* (July 18, 1988), pp. 46-53; William J. Cook (with Jeff Trimble and William Allman), "Red Star Rising," *U.S. News and World Report* (May 16, 1988), pp. 48–54.
10. We were talking at Gary Wegner's home in Hanover, New Hampshire; Bisnovatyi-Kogan and I were both attending an International Astronomical Union Colloquium on White Dwarf Stars in August 1988.

Index